AF710623

HD 9515.6 H64
Hogan, William Thomas
The 1970s

WITHDRAWN

A10202 773464

The 1970s: Critical Years for Steel

The 1970s: Critical Years for Steel

William T. Hogan, S.J.
Fordham University

Lexington Books
D.C. Heath and Company
Lexington, Massachusetts
Toronto London

Copyright © 1972 by D.C. Heath and Company.

All rights reserved. No part of this publication may be reproduced or transmitted in any form or by any means, electronic or mechanical, including photocopy, recording, or any information storage or retrieval system, without permission in writing from the publisher.

Published simultaneously in Canada.

Printed in the United States of America.

International Standard Book Number: 0-669-83063-1

Library of Congress Catalog Card Number: 76-39340

Contents

	List of Tables	vii
	Preface	ix
Chapter 1	**Problems Facing the Steel Industry**	1
	Current Steel Capacity in the United States	1
	Growing Demand for Steel	3
	Steel Profits and Capital Investment	5
	New Criteria for Steel Investment	6
Chapter 2	**Structural Changes and New Technology: Solutions to Steel's Investment Problem**	11
	Diversification in the Steel Industry	11
	Steel Take-Overs by the Conglomerates	16
	Mergers in the Steel Industry	21
	Solutions to Investment Problems	32
	New Steel Technology and Cost Reduction	36
Chapter 3	**Voluntary Restraint Agreements on Steel Imports**	45
	Origins of the Import Problem	46
	Import Penetration of United States Markets	48
	Support for Steel Quota Legislation	52
	Opposition to Steel Quota Legislation	52
	The 1968 Voluntary Restraint Agreements	54
	The 1972 Voluntary Restraint Agreements	62
	Opposition to the 1972 Agreements	71
Chapter 4	**The Steel Labor Agreement and Steel Productivity**	73
	Events Shaping the 1971 Labor Negotiations	73
	Terms of the 1971 Agreement	76
	The Emphasis on Steel Productivity	77
	Output per Man-Hour in Steel	78

	The Concept of Productivity and Its Measurement	81
	Output per Man-Hour as a Productivity Measure	85
	Shortcoming of International Productivity Comparisons	87
	Reasons for the Lag in Steel Productivity	88
	Prospects for Increased Steel Productivity	91
	Plant Productivity Committees	92
Chapter 5	**Factors Affecting Steel Demand in the 1970s**	95
	Construction	96
	Transportation	100
	Durable Goods	107
	Metal Containers	109
Chapter 6	**Conditions Determining the Investment Decision**	113
	Foreign Mergers and Joint Ventures	113
	The Decision to Employ New Technology	116
	Consequences of Noninvestment	119
	Notes	121
	Index	123

List of Tables

1-1	Steel Capital Expenditures in the 1960s	2
1-2	Steel Industry Revenues and Profits, 1962-71 (Millions of Dollars)	6
2-1	Stock Market Performance of Top Eight Steel Companies	17
2-2	Steel Take-Overs by Conglomerate Corporations	18
2-3	Comparative Market Prices, 1969, 1970, 1971	31
3-1	Imports of Iron and Steel Products, 1957-71 (Net Tons)	45
3-2	Imports of Cold-Rolled Sheets, 1964-71 (Net Tons)	48
3-3	United States Foreign Trade in Steel Mill Products, 1957-71 (Thousands of Net Tons)	49
3-4	Comparison of United States Steel Corporation and Bethlehem Steel Corporation Shipments with Import Tonnage, 1967-71 (Millions of Net Tons)	51
3-5	Imports from Nonquota Countries, 1968-71 (Net Tons)	61
3-6	Shipments Received from Quota Countries, 1968-71 (Net Tons)	63
4-1	Growth of Output per Man-Hour in Selected Manufacturing Industries (Average Annual Rates of Growth)	80
4-2	Growth of Output per Man-Hour, Hourly Compensation and Unit Labor Costs for All Employees in Manufacturing, Eleven Countries, 1965-70 (Average Annual Rates of Growth)	81

Preface

Steel is basic to the economy since it supplies the materials for a number of other industries, both large and small. These range from such giants as construction and the automobile industry to small metalworking shops producing a variety of items. In the past few years, despite record sales, shipments, and capital investment, steel has not been particularly profitable. Because of a number of reasons, not the least of which is imports, the rate of return, both on investment and sales, has been low. This raises a vital question about the ability of the industry to justify future investment and attract sufficient capital for the expansion which will be necessary to care for the needs of the economy by the end of the decade. At the present time there is more than adequate capacity for our needs. However, by 1980 more steel will be required.

During the past few months an encouraging trend has developed which indicates that the massive investment made by the industry is beginning to return higher profits. The prospects are that this trend will continue through 1973, and if it can be accelerated and extended beyond that point it will help solve the industry's problems.

This work deals with the principal problems the industry faces, as well as their causes. It also offers some possible solutions.

The author wishes to express thanks to Frank Koelble for his help in the preparation of this study.

The 1970s: Critical Years for Steel

About the Author

William T. Hogan, S. J. is Professor of Economics at Fordham University and Director of the University's Industrial Economics Research Institute. He did his undergraduate work at Fordham College and received both his M.A. and Ph.D. from the Fordham Graduate School.

He has been engaged in economic studies of the iron and steel industry and other basic heavy industries for the past 25 years. He has written many articles during this period, as well as four other books, *Productivity in the Blast Furnace and Open Hearth Segment of the Steel Industry, The Development of American Heavy Industry in the Twentieth Century, Depreciation Policies and Resultant Problems*, and the definitive *Economic History of the Iron and Steel Industry in the United States*, in five volumes, published by Lexington Books.

Father Hogan is a member of President Nixon's Task Force on Business Taxation and a consultant to the Council of Economic Advisors. He has appeared before legislative committees of both houses of the United States Congress including the House Ways and Means Committee and the Senate Finance Committee. He has lectured widely throughout the United States and abroad and been a participant in many conferences on steel industry problems.

1 Problems Facing the Steel Industry

As steelmakers organize their investment planning, the 1970s should logically be a period of expansion. Substantial expenditures will be needed to meet the demand for steel by 1980, but it is difficult to justify the required capital outlay for added capacity on the basis of the low profits recorded in 1970 and 1971. As a consequence, the next few years are critical years for steel, for if the profit picture does not improve materially, there could be a question about future steel investment and about the future growth of the industry.

The steel industry has reached a point of decision in respect to its capital investment plans. Although record sums of money were spent on plant and equipment in the 1960s, additional heavy expenditures are in prospect if capacity is to be kept in line with the growth in steel demand in the years ahead.

By 1980, assuming that the industry expands to accommodate market requirements and that imports are maintained in approximately the current relationship to domestic supply, it is expected that U.S. raw steel production will grow to 175 million net tons. The industry's current raw steel capacity is approximately 165 million net tons, and in order to turn out 175 million net tons on an annual basis, allowing for peak periods of operation, capacity will have to be boosted to between 190 million and 195 million net tons. This will require another massive capital spending effort aimed at the expansion of existing steel plants as well as their continued modernization. The expansion could include the construction of new plants as well as additions to existing facilities. It will also require the acquisition and transportation of additional raw materials needed to sustain the higher level of output.

Current Steel Capacity in the United States

During the 1960s, as indicated in Table 1-1, steel's capital expenditures totaled a record $16.3 billion, 59% more than the $10.3 billion spent in the 1950s. However, despite the sharp rise in expenditures during the 1960s, the net addition to raw steel capacity was far less than in the previous decade. Steel's big expansion in the 1950s saw 48.6 million net tons added to the 100 million net tons that were in place at the start of the decade. This was nearly three times the net addition to total capacity registered in the 1960s, when the rapid installation of BOF (basic oxygen furnace) steel plants and the increased use of electric furnaces occurred simultaneously with a dramatic shutdown of substantial

Table 1-1
Steel Capital Expenditures in the 1960s

Year	Capital Expenditures
1960	$ 1,520.700,000
1961	959,500,000
1962	911,400,000
1963	1,040,000,000
1964	1,599,500,000
1965	1,822,500,000
1966	1,952,700,000
1967	2,145,700,000
1968	2,307,300,000
1969	2,079,200,000
Total	$16,338,500,000

Source: *AISI Annual Statistical Reports for 1961-1970.*

open-hearth capacity and abandonment of the last Bessemer converters. The result was a net addition to total steel capacity of between 16 and 17 million net tons, despite the fact that many more millions of net tons of new steel capacity were put in place.

Unlike the 1950s, which were predominantly years of expansion, much of steel's capital spending in the 1960s was directed at modernization and replacement, with particular emphasis on the widespread change of steelmaking processes. In addition, a substantial investment was made in new finishing facilities, most notably in computerized hot strip mills. Significant amounts of capital also were required to provide for higher investment costs, for diversification into other industries, as well as for the installation of pollution control equipment. This explains in great part the diversity between capital expenditures and the net additions to raw steel capacity during the 1950s as opposed to the 1960s. Comparing the investment accomplishments in both decades, it can be concluded that the industry expanded its overall capacity at a much faster rate in the 1950s and placed much greater emphasis on technological improvements in the 1960s. As a result, it is currently more capable of concentrating its efforts on expansion than it was in the 1960s, when modernization had to assume top priority.

The Revenue Act of 1962 provided for a 7% investment tax credit to spur modernization throughout American industry so that our facilities would be competitive with those in other countries. The steel industry made good use of this incentive to improve much of its plant and equipment. Individual companies must now decide whether or not to embark on a new round of investment in the 1970s in order to keep pace with the expected growth in demand.

The added steel capacity required to bring the total to 195 million tons is only half of the story. The companies will have to replace at least 25 million tons of existing open hearth capacity with either some type of oxygen steelmaking capacity or electric furnaces. Thus if the expansion and replacement job is to be done, at least 55 million tons of melting capacity will have to be installed before 1980.

Growing Demand for Steel

In 1970 a consensus indicated that worldwide raw steel production will reach 1 billion net tons annually by 1980. Of this total, the steel industry in the United States will have to produce about 175 million net tons to meet the needs of our economy. This level of output allows for a 15% share of the market to be absorbed by imports and, depending on the amount of continuous casting in operation, will probably yield 135 to 140 million net tons of finished products for consumption in the public and private sectors of the economy.

In the public sector, large tonnages of structurals and plates will be needed as we go about the business of modernizing or replacing the obsolete and inadequate infrastructure of our economy. A substantial number of bridges will be constructed, some as new links in our growing transportation system while others will replace outdated spans built in the early part of the century. The highway system must also be expanded, with substantial provision made for overpasses and cloverleafs to eliminate traffic bottlenecks. In addition, public buildings such as schools and hospitals will have to be replaced and expanded.

In the private sector of the economy, with a population of 237 to 240 million expected by 1980, the demand for consumer durable goods, such as automobiles, appliances, and household furnishings will also expand. Further, over 20 million housing units must be constructed, and these will include not only traditional stationary homes, but also a sizable number of factory-built and mobile homes. The container market should also experience a substantial growth, with steel's participation enhanced by the development of tin-free can stock and new processes for producing drawn and ironed two-piece cans and for shaping cylindrical cans into product-related shapes, such as miniature beer kegs. As a result of the new shaping process, one brewer has already moved to replace 25 million aluminum cans annually with steel.

One of the new steel markets starting to open up is tied to our growing requirements for energy, which have increased appreciably and are continuing to expand. In the face of this growing demand, gas reserves of the United States have started to decline; at the end of 1970, they were approximately 291 trillion cubic feet, and by December 31, 1971, they had dropped to 279 trillion cubic feet. Production last year was in the area of 22 trillion cubic feet, and if this figure remains constant without any immediate prospects of discovering new reserves, it will exhaust our supply in a relatively short time. Therefore, it is imperative that we import gas in some form or other, and one of the most likely alternatives at present is liquified natural gas.

The plant and equipment, as well as the ocean vessels, required to carry out a large-scale transportation of liquified natural gas, will require an investment of hundreds of millions of dollars. The tanker fleet alone, which could reach as many as 180 to 200 vessels on a worldwide basis by 1980, will require a much larger investment than the conventional oil tanker. The prospect is that many of these will be built in the United States, particularly if subsidies are granted under the Merchant Marine Act of 1970. The cost will be between $85 and $95 million per vessel. Each ship, if it is to carry 125,000 cubic meters of LNG, will consume approximately 35,000 tons of steel, the predominant items being structural steel and plates. Further, there is one type of carrier which could use as much as 4000 to 5000 tons of some form of stainless steel for the special tanks to hold the liquified gas. The terminal facilities and massive storage tanks needed for handling LNG at the receiving port will also use large tonnages of steel. Further, when the LNG is gasified, extensive new pipeline systems will be needed for distribution to the consumer.

A sizable new market for stainless steel is starting to emerge as the automobile industry prepares to meet emission control standards set down by the Environmental Protection Agency for 1975 and 1976. Control systems under consideration could use as much as 65 to 70 pounds of stainless steel per vehicle, and with a production of 9 million cars and 1.5 to 2 million trucks, this could mean a market in the area of 400,000 tons annually. Further, the new, collision-resistant bumpers to appear on 1973 models will increase a car's weight by 40 to 100 pounds, with much of the addition consisting of steel. Auto safety programs have already seen the widespread introduction of side-guard door beams, and the experimental safety vehicles now in testing use substantially more steel in their current production models in such new body parts as roll bars, structural bumpers, engine and transmission support beams, and the chassis beam assembly. Finally, if the auto industry gets around to a major conversion to the Wankel engine, the retooling involved will create a substantial demand for steel.

The installation of air and water pollution control facilities will also provide a large market for steel, both carbon and stainless. In industries such as petroleum, chemicals, pulp and paper, utilities, and steel itself, it is estimated that in many cases up to 10% of capital expenditures for new facilities will be devoted to air and water pollution control. Further, existing facilities must be equipped with pollution controls and these will be costly. In many instances, the actual control facility will cost more if installed on an existing piece of equipment than engineered into a new facility. Huge ductwork systems are required to control the flow of emissions. In some cases, particularly in the steel industry, the ducts are as much as 20 feet in diameter and must be fabricated from plates. In one instance, just one section of a ductwork system was 20 feet 9 inches in diameter, 95 feet long and weighed approximately 45 tons.

While the pollution control facilities do not improve the steelmaking process in terms of quantity or quality, they do in themselves contain a great deal of steel and should provide a sizable market throughout the rest of the decade. Furthermore, these are not postponable expenditures; they are necessary in all industries if air and water pollution control specifications are to be met. Thus the steel industry will find a market in pollution control facilities. However, in this case it will probably be one of its own best customers.

Steel Profits and Capital Investment

The additional capacity over and above what is necessary for replacement and modernization will require not only 25 million to 30 million tons of steel melting equipment, but also the supporting facilities which, in a number of cases, will be blast furnaces and coke ovens. In addition, rolling and finishing facilities will also be required for certain products such as plates and structurals.

It is doubtful whether any additional hot strip mill capacity will have to be provided, although there is certainly need for the replacement of a number of hot strip mills. To justify the installation for the needed expansion, profits will have to be much higher in 1972 and 1973 than they were in the two prior years.

Profits for 1970 and 1971 amounted to a return of 2.8% on sales in each year and a return on stockholders' equity of 4.1% in 1970 and 4.3% in 1971. These figures were posted despite the fact that total revenues for the industry, with the exception of 1967, showed continued growth from 1962 to 1971 when they surpassed $20 billion. Despite this, profits have declined markedly from the 1966 high of $1.075 billion to a low $531.6 million in 1970, with a slight recovery to $566.2 million in 1971.

Table 1-2 gives the revenue and profit picture from 1962 to 1971, as well as the percentage return on revenues and stockholders' equity.

In order to justify an investment in new facilities for expansion, the profit picture must improve substantially in the next two years. This is why 1972 and 1973 are truly critical years for the steel industry. The decision will have to be made by the end of 1974 if the new facilities are to be operative in 1977, 1978, and 1979, for it takes two or three years from the time a decision is made and the engineering begins until the equipment is actually at productive capacity. Thus, if decisions are postponed beyond early 1975, it is doubtful that the capacity will be available when it is needed in the late 1970s. The amount of investment necessary for the additional capacity as well as for normal replacement and pollution control will most certainly be in excess of $1.7 billion a year. Just what percentage return on investment and sales will be necessary to bring forth these dollars is hard to determine. It certainly will have to be almost double the figures posted in 1970 and 1971.

Table 1-2
Steel Industry Revenues and Profits, 1962-71 (Millions of Dollars)

Year	Revenue	Net Income	% Return on Revenues	% Return on Equity
1962	$13,980.6	$ 566.4	4.1	5.3
1963	14,612.6	782.0	5.4	7.1
1964	16,357.1	992.3	6.1	8.7
1965	17,971.7	1,069.3	5.9	8.9
1966	18,288.4	1,075.3	5.9	8.9
1967	16,880.4	829.8	4.9	6.8
1968	18,679.6	992.2	5.3	7.9
1969	19,231.0	879.4	4.6	6.9
1970	19,269.5	531.6	2.8	4.1
1971	20,126.2	566.2	2.8	4.3

Source: *AISI Annual Statistical Report for 1971*, p. 10.

New Criteria for Steel Investment

A new round of heavy investment in steel, similar in planning and execution to that which occurred in the 1960s, cannot be justified under current economic circumstances. A viable course for plant and equipment spending by steel companies over the remainder of the 1970s must recognize:

1. That steel has become an international commodity produced and sold by large, restructured overseas companies, who with government approval and cooperation, jointly plan investment and production programs and establish pricing and marketing policies
2. That in certain segments of the domestic industry, recent capital programs have resulted in capacity additions in excess of demand requirements and have placed melting and finishing capabilities out of balance
3. That increased fixed-cost burdens and worsening internal financial positions in the industry have adversely affected the availability of capital for steel investment
4. That investment costs have been increased substantially by pollution control requirements

Total capital expenditures in steel during 1960-70 inclusive totaled over $18 billion and averaged about $2 billion annually during the five-year period 1966-70. During these years, the growth of foreign competition increased the pressure to invest in new technology in the interest of improved operating efficiency and high product quality. However, while steelmakers in the United States made record investment commitments to install basic oxygen converters, continuous casters, and the most advanced finishing facilities, equivalent technology was available overseas, and its adoption both in Europe and Japan was pursued in a disciplined, rationalized manner by very large producers, newly structured by a series of steel company regroupings which made them capable of a more intensive use of large-scale facilities. In addition, domestic costs for plant and equipment, labor and materials were substantially higher than those prevailing abroad, and consequently U.S. producers—despite record capital outlays—remained noncompetitive on a price basis. Ultimately, foreign producers became firmly entrenched in the domestic market, so much so that their tonnage shipments here in 1971 exceeded those of Bethlehem Steel Corporation, the nation's second largest producer, by 5.7 million net tons and were only 1 million net tons less than the shipments of United States Steel Corporation.

The inability of massive infusions of capital to stem the import tide was nowhere more apparent than in the light, flat rolled products segment of the steel business which traditionally accounts for the dominant share (46-48%) of total industry shipments. During the 1960s, domestic producers invested heavily in modern hot strip mills and other related equipment to strengthen their hold on this most vital of all steel markets. So superior were these new computer-controlled rolling facilities to their predecessors, even those constructed in the 1950s, that they rightfully were termed a "new generation" of hot strip mills. They operate with much greater speed and precision, yield a higher quality product more uniform in gage, and accommodate much larger slabs, making it possible to roll much larger continuous coils (e.g., 72 to 82 inch wide coils weighing 1100 lbs. to 1250 lbs. per inch of width, or more than 40 tons, about double the maximum coil weights on prewar strip mills). In effect, the new mills provide the wherewithall to meet customer demands for larger coils and closer strip tolerances and, consequently, their output is much more competitive than that turned out on older equipment. This made "second generation" hot strip mill capacity a prerequisite to compete effectively in the light, flat rolled products market upon which steel companies depend for a major source of their revenues.

Despite the tremendous capital costs involved (i.e., from $80 to $135 million per mill, according to the size and complexity of the facility installed), ten U.S. companies found it necessary to make such investments, not only to meet domestic competition, but also to curtail the loss of business to Japanese and European producers. Nevertheless, with modern facilities also installed overseas, flat rolled product imports increased rapidly. In the early years of the decade, 1960-62 inclusive, import penetration amounted to a minimal 1.1% share of the combined domestic consumption of sheet, strip, and tin plate. In 1971, however, 8,371,537 net tons of these products were imported, nearly 23 times the 1960-62 average and a distressing 16.8% of the domestic market.

The massive investment in modern hot strip mills, while unable to head off the influx of imports, resulted in an inordinate expansion of U.S. hot rolled sheet capacity and the subsequent underutilization of costly new facilities. At the same time it contributed to a worsening of steel company debt-equity relationships. This combination of conditions, in turn, has helped to seriously depress rates of return on capital invested in the steel business.

Between 1960 and 1970, U.S. hot strip mill capacity increased by two-thirds, from 53,000,000 net tons per year to 88,000,000 net tons. Over this same period, however, domestic shipmens of sheet, strip, and tin mill products increased only 20.84% from 35,042,000 to 42,344,000 net tons. The imbalance is based on conservatively fixed current capacities for the new mills installed since 1960, which taken together are rated at 34,422,000 net tons. However, prescinding from the operating limitations currently imposed by outside factors, the combined potential capacity of these mills is much greater, and the excess of capacity over demand is magnified. With proper scheduling and the availability of steel plus the right size slabs in adequate tonnages, each second generation mill should be able to produce 4.5 million to 5 million net tons per year. This would result in a 17 to 23 million net ton per year increase in their combined current capacity of 32,422,000 net tons, and would boost total U.S. hot strip mill capacity significantly above the 100 million net ton level. Admittedly, this is only a theoretical capacity level, since a number of the older hot strip mills currently in operation will be phased out as recently installed facilities work toward their operating limits; but it does serve to underscore the problems of duplicated steel investment and short-term overcapacity.

The financial position of the steel industry has deteriorated considerably in the last decade; not only have its profits had a downward trend since the mid-1960s, but other financial indicators have followed suit.

During 1965-69, while the industry's invested capital, considered on an average annual basis, increased 47.6% above the level of 1955-59, its tonnage steel shipments and net sales volume, both restricted by the influx of imports, rose by only 17.1% and 26.6% respectively. With capital expenditures sustained at extremely high levels throughout the 1960s, the industry's reliance on debt-financing increased sharply. During 1965-69, its long-term debt jumped 116.9% above the average annual level of 1955-59, while its stockholders' equity rose by only 33.6%.

With the underutilization of expensive new facilities, indicated by the disproportionate growth of invested capital and steel shipments, as well as the sharp rise in the industry's debt burden which has seen its annual interest payments increase from $101.4 million in 1960 to $323.2 million last year, the fixed costs of steel production have risen significantly. At the same time, the industry has incurred a variety of operating cost increases, such as those for labor, materials, and transportation, resulting in a marked downward trend in profits.

In recent years, steel profits have been adversely affected by industry-wide pollution control programs which have increased capital costs and manpower requirements without any corresponding increase in steel output. Spending by steel companies to control air and water pollution has accounted for a growing share of their total capital expenditures, and to date such spending has aggregated over $1.3 billion, more than in any other major industry. If provision for pollution control is made in the design of a new, integrated steel plant constructed from the ground up, its cost currently represents in some cases as much as 10% of the plant's total required investment. However, it is generally more costly to install pollution controls at older plants, given the resulting restriction of production, the problem of providing adequate space, and the need to reengineer and modify existing equipment and structures. During the next three to four years, steel companies will have to invest in excess of $3 billion in air and water pollution controls if they are to comply fully with air and water quality standards. This expenditure will constitute the number one priority item in the industry's investment planning.

As individual steel companies make their investment plans for the remainder of the 1970s, they are confronted with the serious obstacles of sharply escalated investment costs on the one hand and unsatisfactory levels of productivity and profits on the other. As a first step to justifying additional capital spending, the industry must attempt to ease the impact of imports, which took somewhat over 17% of the market in 1971. It must also, by increasing productivity wherever possible, strive to keep its unit labor costs down. If improvement can be managed in these two areas, the results will be higher volume, lower costs, and ultimately, higher profits. However, improved returns on steel investment and improved capital availability cannot, in themselves, support an affirmative decision on future steel spending for expansion. The industry cannot afford more of what occurred in the 1960s; rather, it must seek to establish a more reasoned approach to investment, one that permits needed capacity to be installed without wasteful duplication and capacity imbalances and at the same time takes advantage of the lowest-cost, most efficient, most pollution-free alternatives for meeting increased demand in the years ahead.

Thus far in 1972 there have been signs of improvement and recovery. If these result in substantially increased profits both in 1972 and 1973, there is a very strong possibility that an affirmative decision can be made on investment in steel expansion for the remainder of the decade.

In terms of planning and execution, the industry's approach to future spending, given the evolution of steel as an international commodity, should be more nearly comparable to that pursued by its competitors overseas. This requires that steel companies be permitted to jointly participate in capital projects and that investment be further rationalized on an industry-wide level, much in the same manner as in Europe and Japan. Such an approach to investment could result in some revision of the industry's structure, possibly involving mergers to permit a fuller utilization of capital.

2 Structural Changes and New Technology: Solutions to Steel's Investment Problem

During the remainder of the 1970s, the amount of capital spending by steel companies to meet the growth of market demand may well be influenced by the extent to which joint ventures and mergers can be utilized to streamline traditional investment practices. In the past, joint ventures have been prevalent in only one segment of the steel business, namely the production of iron ore, while the vast majority of steel company spending has involved individual decisions to install plant and equipment throughout every other phase of the steelmaking process. Under this practice of massive spending on a company by company basis, steel mill investment has become a high-cost-low-return venture, less capable of competing for essential capital, whether for modernization or expansion. Consequently, at the present time the need to attract capital for the installation of additional steel capacity requires a new approach to spending, one which could benefit significantly from joint ventures and possibly mergers, as well as from the continued use of new and emerging technological advances with a potential for lowering capital costs.

In gauging the prospects for future change in the structure of the steel industry, it is important to consider that a widespread reorganization has been in progress since the early 1960s, and it appears that the end is not yet in sight. The result has been fundamental corporate change, which in many of its aspects reveals a favorable disposition within the industry to the adoption of new business and investment strategies. The frequency of this change and the variety of directions it has taken represent a thorough departure from past experience. During a period of more than sixty years following the formation of U.S. Steel Corporation in 1901, the basic structure of the industry—although modified on occasion—remained essentially intact. By contrast, recent years have seen the industry restructured from within by means of steel company diversification and intraindustry mergers and simultaneously from without through mini-steel plant development and steel take-overs by conglomerate corporations. The magnitude of the current reorganization is indicated by the fact that intraindustry mergers and conglomerate take-overs have already involved steel companies with annual sales currently in excess of $6.0 billion.

Diversification in the Steel Industry

In the middle to late 1960s, diversification into nonsteel businesses became one of the foremost instruments for changing the structure of the steel industry. Its use to level out the ups and downs of the steel cycle had long been discussed and contemplated in industry circles, particularly during periods of slack demand, and over the years a number of steel companies developed interests in other fields. It was not until 1966-67, however, that an industry-wide movement into diverse forms of enterprise started to take shape, reaching its peak toward the end of the decade and slowing down appreciably in the early 1970s.

In contrast to the post-1966 diversification movement, which was extremely aggressive and effected fundamental revisions in corporate structure, many of the early excursions into nonsteel activities were treated as outside interests. The idea of diverting company resources from an all out effort on the steel front, whether to insure against the effects of cyclical swings or for any other reason, never had much appeal for steelmakers. Rather, periods of recovery in the steel cycle were counted on to generate activity and profits sufficient to compensate for the disadvantages endured when business turned downward. However, as other competitive elements captured an increasing share of the steel market, the industry's profit picture, even during good times, compared less and less favorably with that of other industries. This served to enhance the prospects for more profitable investment outside of steel and increased the incentive to diversify. By 1966-67 virtually every large steel company had either acquired a nonsteel interest or had formed a committee to actively investigate nonsteel projects.

United States Steel Corporation

United States Steel Corporation which had for years been engaged in such fields as cement production, coke oven by-products and shipping, made a strong move into the chemical industry in 1964 with the purchase of the assets of the Industrial Chemicals and Protective Coating Division of Pittsburgh Chemical Company. The corporation explained its move in its *1964 Annual Report*:

> This division produces basic materials for the plastic industry, tar-based enamels for pipelines, tar and asphalt base coatings and other industrial chemicals. In the manufacture of many of these products a number of coal chemicals are used as raw materials, and United States Steel is an important producer of such coal chemicals.

United States Steel Corporation's Annual Report for 1966 noted that the USS Chemicals Division was formed in February of that year to organize the production and marketing of chemicals, primarily those recovered from coke oven gas. The *1966 Annual Report* also noted that:

> This development represented 'a basic diversification and a logical step arising

from U.S. Steel's established and long-term position as a major producer of chemical raw materials.'

In mid-January of 1968 USS Chemicals announced plans to double its production of melic anhydrite and substantially increase its output of fumoric acid. A new production complex for these chemicals was built at the Neville Island (Pennsylvania) Plant and began operations in 1969. Some of the chemicals produced are used as raw materials to make plastics as the United States Steel Corporation became a plastics producer through the purchase of the plastics division of General American Transportation Corporation. This new operation consisted of five plants located in four states.

There were also ventures into titanium, financing services, engineering and consulting, as well as real estate. This move to diversification was highlighted by a statement on the cover of the *1968 Annual Report* which described the United States Steel Corporation as "A Diversified Producer of Materials and Services."

A further indication of United States Steel's intent to diversify was given in a statement by Edgar Speer in April of 1969, shortly after he became president of United States Steel Corporation. When asked a question at a press conference: "What does U.S. Steel intend to do to combat the use of plastics by the automobile industry?" Mr. Speer simply said, "Go into the plastics business."

Bethlehem Steel Corporation

Bethlehem Steel Corporation made an attempt in 1967 to acquire the Cerro Corporation which was engaged in the mining and production of copper and other nonferrous metals. This was a substantial company with sales in excess of $400 million annually. However, the Justice Department gave Bethlehem an adverse ruling and the matter was dropped. Bethlehem has since acquired Kusan, a producer of plastic parts, as well as an eighty percent interest in Multicon, a producer of houses, and thus has begun to diversify its operations.

National Steel Corporation

National Steel in 1963 acquired two large steel service center companies, namely, Korhumel Steel & Aluminum Company and A.R. Purdy Company. This expanded the number of outlets from one in Houston to ten located in the midwest and eastern part of the country. In the strictest sense, entrance on a large scale into the service center segment of the steel industry was not a diversification. However, it did change the structure of the company to some extent.

In 1968 National Steel entered the aluminum industry by becoming a joint owner of the fifth largest primary producer of aluminum. In May of that year it acquired 20% of the South Wire Company of Carrollton, Georgia, and along with this company built a primary aluminum reduction plant at Hawesville, Kentucky. It also acquired Republic Foil of Danbury, Connecticut, and Hastings Aluminum Products of Hastings, Michigan, as well as the Ward Steel Company with service centers in three locations as an outlet not only for steel, but aluminum products as well. National also added Howe Engineering and Ward Supply whose facilities were to be used to fabricate aluminum products. Thus the company had a fully integrated aluminum operation starting with smelting and proceeding through rolling to fabrication and distribution facilities. The investment in aluminum was in excess of $100 million.

In 1969 National also entered the magnesium industry as a result of 30% participation in the American Magnesium Company which had a primary magnesium plant at Snyder, Texas.

Armco Steel Corporation

In the Armco Steel Corporation *1966 Annual Report*, the letter of the president and chairman announced the decision to pursue nonsteel activities.

> GROWTH OPPORTUNITIES PURSUED
> We are aggressively exploring growth opportunities in the production of other materials which serve the markets we believe have the greatest growth and profit potential. The decision last December to acquire 29 percent interest in a titanium company is illustrative of this activity.
>
> To pursue other growth opportunities, Armco has established a new Strategic Planning Task Force composed of members of executive management. The mission of this group will be to study thoroughly new markets, products and processes, here and abroad, on which Armco's human and material resources can logically be brought to bear, in a world in which accelerating technology can influence a company's operations so profoundly, a company which doesn't seek out new ideas and new visions cannot grow for the maximum benefit of its shareholders.

In keeping with its newly stated objective, Armco organized the Advanced Material Division in February of 1968. Its function was to concentrate on the production and sales of stainless steel, super alloys, and titanium. The *1968 Annual Report* devoted an entire section to diversification.

Early in 1969, the company organized the Armco/Boothe Corporation which established a partnership with GATX/Boothe and thus entered the field of finance leasing. Its objective was to lease industrial machinery, ships, railroad rolling stock, jet aircraft, and other capital equipment. Armco/Boothe planned to place $150 million worth of equipment under lease in 1969.

Late in the same year, Armco merged with HITCO. This company was one of the largest producers of nonmetallic composites, many of which were used for components in aircraft, space vehicles and rockets. The company also manufactured a variety of other products including cloth, yarn, tape filaments, and molding compounds.

Inland Steel Company

In 1968 Inland Steel move to diversify into nonsteel operations. The *Annual Report* for that year stated:

> Inland has adopted a more aggressive diversification policy. To implement this policy a corporate development staff was formed last May to develop new investment opportunities.
>
> Inland is well prepared to meet the anticipated growth in steel demand, in terms of production capacity, product quality and customer service. The new diversification program will enable us to benefit as well from the rapidly growing demand for other products, services and materials.
>
> Chief criteria to be applied to possible acquisition are rate of growth and profitability.

Under its new policy Inland entered into a joint venture with A.O. Smith Company for the production of powdered metal, strong lightweight, corrosion-resistant reinforced plastic pipe, and molded reinforced plastic parts.

In 1969 Inland acquired a 30% interest in On-Line Decisions, Inc., a computer software company located in California. In the same year, the company acquired Scholz Homes, Inc., of Toledo, Ohio. This new acquisition gave Inland a substantial foothold in prefabricated housing, mobile homes, sectional housing, single family housing, and apartment buildings.

In order to supplement the benefits to be gained from the acquisition of Scholz Homes, Inc., Inland formed the Inland Steel Urban Development Corporation (INSTUD). Through another unit, the Inland Steel Development Corporation, the company participated in the purchase of a large tract of land along the Potomac River for the purposes of a housing development.

Inland hopes to obtain approximately one-third of its income through the operation of these nonsteel activities.

Republic Steel Corporation

In 1969 Republic Steel Corporation organized a subsidiary, Republic Steel Enterprises, Inc., in order to investigate the possibility of entering nonsteel activities. Among these there were the real estate and housing fields. Republic Steel Corporation had considerable acreage of land in various parts of the country, including areas at Birmingham, Alabama and Cleveland, Ohio. To foster the development of this land into industrial and commercial uses, Republic Steel organized a consortium of six companies with experience in commercial, industrial, and residential development.

These are but a few examples of the trend to diversification. It should be noted that the nonsteel efforts of large steel companies have to be substantial in size to make them worthwhile. One major steel company stated that it would not be interested in any enterprise with less than $50 million of sales because a small volume of sales would hardly be profitable, particularly when the current total sales of that company are in excess of $2 billion.

As has been noted, the trend toward diversification has slowed up considerably in the past year or two. This was due to the fact that some of the ventures were successful while others were not, and the industry has paused to assimilate the new ventures.

Steel Take-Overs by the Conglomerates

The accelerated pace of diversification in steel during the late 1960s, typified by the extensive programs already detailed, can be attributed to a great extent to the above stated factors, however one additional item is significant in explaining the trend. This was the abrupt outburst of another form of corporate change—the steel take-over. Understandably, steelmakers preferred to put their own assets to work and were not too appreciative of the offers made by conglomerate managers to do it for them. Nevertheless, the attraction of steel company assets and low stock market prices for steel equities resulted in a long list of take-over bids by the conglomerates, and in 1968 the steel take-over became the most obvious and controversial aspect of the industry's reorganization.

Steel stocks were out of favor with Wall Street during most of the 1960s. As a group they reached a peak in 1959 and backed off through 1961, when steel output and sales declined and costs increased significantly. Then in 1962, the steel price rollback forced by the Kennedy administration helped to establish a stock market attitude leading to a long-term depression in the performance of steel equities. Year-end 1967 stock prices for the top eight steel producers (see Table 2-1) ranged from 8.4 to 12.8 times earnings, substantially below the multiples of 10 to 25 times per share earnings which prevailed during the 1959-62 period. This downgrading of the steel stocks after 1959 occurred despite new record tonnages of production and shipments, and the tendency to overlook a number of basic values eventually gave rise to a growing interest in steel companies as take-over candidates. The high cash flows and low price-earnings ratios which failed to impress investors eventually caught the eye of acquisition minded conglomerate corporations. If Wall Street wasn't buying the steels, the conglomerates were.

Table 2-1
Stock Market Performance of Top Eight Steel Companies

Company	Highest Price	Year High was Reached	Closing Year End 1967	% Decline
U.S. Steel	108 7/8	1959	40 3/4	62.57
Bethlehem	59 1/8	1959	32 5/8	44.82
Republic	81 3/4	1959	43 5/8	46.64
Armco	80 1/2	1959	48 1/2	39.75
National	65 3/4	1965	46	30.04
Inland	55	1959	33 3/4	38.64
Jones & Laughlin	89 3/4	1960	55 3/4	37.88
Youngstown Sheet and Tube	55	1964	31 1/4	43.18

During 1968, five out of the top twenty companies in the steel industry were the object of take-over bids by conglomerate corporations. They included Jones & Laughlin Steel Corporation and Youngstown Sheet and Tube Company, the nation's seventh and eighth largest steel producers, Crucible Steel which ranked number twelve in the industry overall and was the nation's second largest producer of specialty steels, as well as CF & I Steel and Sharon Steel which ranked fourteenth and sixteenth in the industry. The steel companies taken over by conglomerate corporations are listed in Table 2-2.

Actually many more outsiders were involved in steel take-over bids than the scorecard indicates. This is brought to light by the events surrounding a number of the match-ups listed above.

Youngstown-Lykes. Prior to Youngstown Sheet and Tube's merger with Lykes Corporation, a New Orleans based holding company with large steamship interests, the steelmaker had been approached by Ling-Temco-Vought, and Signal Corporation and following the move by Lykes held merger talks with Avnet, Inc., a diversified New York concern producing electric components. Youngstown and Avnet had actually announced their intention to merge, but were dissuaded from seeking the required two-thirds affirmative vote by Youngstown stockholders when Lykes announced its ownership of almost one-fourth of the stock.

Crucible-Colt. Crucible Steel, which has become a subsidiary of many faceted Colt Industries, had been sought by two other conglomerates, Walter Kidde and Company, Inc., and Whittaker Corporation.

Sharon-NVF. Sharon Steel, despite the opposition of its management, was taken over by the NVF Company, a producer of industrial plastics, vulcanized fibre,

Table 2-2
Steel Take-Overs by Conglomerate Corporations

Take-Over Targets	Company Making Offer	Year Offer Initiated
Jones & Laughlin Steel	Ling-Temco-Vought	1968
Youngstown Sheet and Tube	Lykes	1968
Crucible Steel	Colt Industries	1968
CF & I	Crane	1968
Sharon Steel	NVF	1968
Lone Star Steel[a]	Northwest Industries	1965
Jessop Steel	Athlone	1968
Firth Sterling	Teledyne	1968
Columbia Steel	Teledyne	1968
Continental Steel	Penn-Dixie Cement	1968

[a]First acquired by Philadelphia Reading and in turn was acquired by Northwest Industries.

paper and containers. NVF won out over two larger suitors, Alloys Unlimited, Inc., and Cyclops Corporation, which had Sharon's cooperation in making their bids. It should be noted that the overtures by Cyclops were opposed by the Justice Department.

Jessop-Athlone. Prior to its acquisition by diversified Athlone Industries, more than passing interest in Jessop Steel had been shown by no less than five companies.

To be sure, the steelmakers which have been taken over rarely had an opportunity to hand pick their new partners, and negotiations preceeding acquisition usually did not occur under the most cordial or cooperative of circumstances. On the other hand, where contests for shareholder support and competitive bids occurred, offers were sweetened to the benefit of the steel companies involved. The most notable exception to competitive bidding occurred in the case of Jones & Laughlin, but in this instance, the cash offer from LTV was such that it was not about to be matched by another bidder nor refused by the steelmaker. LTV's \$85 per share cash offer aggregated \$425 million, the largest cash tender ever made, and its attractiveness produced an 86% response to an initial offer to purchase 63% of Jones & Laughlin's outstanding common shares.

In contrast to LTV's straight cash tender, transactions in other steel take-overs involved various combinations of cash, common and preferred stock, warrants and debt securities. It should be noted that the debt securities issued in take-overs afford tax advantages to the acquiring concern and to the stockholders of the company acquired. Unlike common stock dividends, which are drawn from after-tax profit, interest on the debt is tax deductible as a corporate business expense, and any capital-gain tax paid by stockholders on the exchange of their common stock for debentures is deferred until the principal is repaid.

In the Youngstown-Lykes hookup, the steel company's shareholders were offered a new Lykes convertible preferred and the opportunity to exchange their shares for debentures, whereas NVF's take-over of Sharon Steel, a company more than seven times its size based on both sales and assets, was accomplished with the package of debentures and warrants to purchase NVF common stock. In contesting with other bidders for the shares of their respective steel targets, both Lykes and NVF sweetened their offers by increasing the principal amount of their debentures. NVF also made its offer more attractive by adjusting the terms for the exercise of its warrants. In the Crucible-Colt consolidation, a new corporation was formed under the Colt name, and Crucible shareholders received fractional shares of both its common stock and a convertible preferred in exchange for their Crucible common.

The acquisition of Continental Steel Corporation, Kokomo, Indiana by the Penn-Dixie Cement Corporation was 53% completed as of August, 1972, and intentions have been announced to acquire the remaining 47% interest by issuing debentures and stock valued at about $20.4 million. The transaction, calling for the eventual merger of Continental into Penn-Dixie, should be finalized by the end of this year.

Unlike the Continental acquisition, which is still in progress, most of the steel takeovers initiated in 1968 were completed in 1969, just prior to the downturn in steel activity during 1970-71. In many instances, the cash flows that had attracted the acquisition-minded conglomerates fell far short of expectations, and the financial difficulties that ensued, coupled with a Justice Department move to regulate conglomerate mergers, effectively dampened the takeover fever, not only in steel but in other industries.

To a great extent, the slowdown of the conglomerate movement in recent years can be traced to the series of events set in motion by LTV's acquisition of Jones & Laughlin. On May 1, 1968, Jones & Laughlin's common stock closed at 49 1/8. By May 9, it reached 62 and was 77 on May 10, when trading of both the common and preferred issues was suspended by the New York Stock Exchange. On May 11, LTV offered to buy a 63% interest for cash at $85 per share and stated that the Jones & Laughlin management would not oppose the offer. On June 25, LTV announced that it had acquired 5,001,548 shares for slightly more than $425 million. The conglomerate's own funds provided $200 million of this huge sum, and the remainder was borrowed from banks and institutions, with $100 million coming from Canadian and European sources.

This takeover of the nation's seventh largest steel producer with more than $1 billion in annual sales by a $2.8 billion enterprise with interests in aerospace, air transportation, meat packing, pharmaceuticals, wire and cables, sporting goods, and electronics fostered a growing opposition to the conglomerate movement, and on May 24, 1969, the Justice Department moved to test the applicability of the antitrust laws to conglomerate mergers by instituting a suit to force LTV to dispose of its Jones & Laughlin holdings. Three days later, LTV and the Justice Department announced an agreement under which LTV was allowed to buy up to 81% of Jones & Laughlin's stock, provided that the steel producer would be permitted to operate as an independent company and provided that LTV would agree to divest itself of all interests in Jones & Laughlin if the Justice Department won its case.

As 1969 progressed, changes in the economy started to affect both the conglomerate and the steelmaker. LTV, which had earned $33.9 million in 1967 and $27.7 million in 1968, lost $6.5 million in the first quarter of 1969. An additional loss of $3.3 million in the second quarter contrasted with a net profit of $10.6 million for the comparable quarter of 1968. By August 1969, LTV was required to pay 9.5% interest to obtain an extension of a $100 million loan from twenty-five major banks. The Jones & Laughlin purchase had brought LTV's total long-term debt to $865 million.

Jones & Laughlin had lost $185,000 during the last quarter of 1968. In 1969 it earned $22.1 million (down from $65.1 in 1966) and paid out $22.3 million in dividends. The steady slide in Jones & Laughlin's earnings after 1966 cast some doubt on its ability to help carry the immense debt LTV had assumed in its purchase. The danger became real in the first quarter of 1970, when the steelmaker lost $1.2 million in comparison to a net profit of $9.9 million for the like period in 1969. In spite of this, a dividend was paid for the first quarter, but the second quarter dividend was passed for the first time in twenty-nine years.

The financial difficulties surrounding LTV's takeover of Jones & Laughlin did not go unnoticed when the sharp break in the stock market occurred in 1970. LTV's common dropped to 7 1/8 in contrast to its high of 169 1/2 in 1967. On June 10, 1970, a settlement of the Justice Department's suit against the take-over was stipulated by Federal Judge Louis Rosenberg. In order to retain its 81% interest in Jones & Laughlin, LTV was required to relinquish its wholly owned cable and wire subsidiary, Okonite Company and its 55.4% interest in Braniff Airways. Further, LTV was restricted in the operation of the Jones & Laughlin employee pension or other benefit funds, and any change in the operation of these funds had to be approved by the Justice Department and Judge Rosenberg's court.

The test case challenging LTV's acquisition of Jones & Laughlin was interesting insofar as it called for a reinterpretation of the antitrust law to reflect "new" forms of corporate enterprise. Prior to the change of administrations in Washington in 1969, the accepted role of antitrust policy had developed principally along the lines of keeping a close check on combinations between firms in the same or closely related industries, while conglomerate mergers between totally unrelated companies were considered outside of the government's jurisdiction. Within this regulatory framework, 85% of the 200 or so mergers concluded in 1968 between companies with assets exceeding $10 million were of the conglomerate type. In effect, the law and its implementation had been instrumental in creating the conglomerate movement. The long-standing tendency to oppose mergers between companies in like businesses had served to discourage many such combinations, which would have made good economic sense, and conversely, gave rise by default to mergers between companies in widely-scattered businesses.

During the post-World War II period, the blending of traditional lines of commerce ran counter to the government's norm of antitrust regulation, which sought to establish the degree of competition or concentration within particular industrial categories set by the Census of Manufacturers. This approach made the underlying assumption that competition is confined within specific industries and that it varies in direct proportion to the number of firms listed in each of the Census of Manufacturers' groupings. In so doing, it failed to give adequate recognition during the 1950s to interindustry competition derived from accentuated trends of substitute product development and industrial diversification. Ten years ago there was already a substantial list of conglomerates that defied classification into any single industrial category.

The preponderance of steel take-overs by conglomerates versus mergers between steel companies as an instrument for structural change in the industry undoubtedly was a reflection of the government's policy on business combinations. Given a more realistic approach to the evaluation of concentration and competition, there would have been a greater inclination within the industry to steel company mergers. This would have been particularly opportune during the late 1950s and early 1960s when small steel companies, both integrated and nonintegrated, met with considerable difficulty in raising adequate capital to install new production techniques and equipment. At that time, a careful pooling of the resources of these companies could have created a number of financially-sound, technically viable competitive entities in the steel business. Their formation could have eliminated obsolete equipment and significantly cut overhead costs, in addition to providing a diversity of steel product mix and more ready access to the capital market. Few of the recent steel take-overs can be justified on equivalent economic grounds.

Mergers in the Steel Industry

During the past decade there have been four major mergers in the steel industry. Each involved two companies which, in every case, complemented the needs of both. The first took place in 1964 between the Interlake Iron Corporation and the Acme Steel Company. The second was between Pittsburgh Steel Company and Wheeling Steel Corporation in 1968. The third was between Cyclops Incorporated and Detroit Steel Corporation in 1970 and the fourth was between the National Steel Corporation and the Granite City Steel Company in 1971.

Acme-Interlake

This was the first major merger in the steel industry during the postwar period. It was consummated in December 1964 and resulted in a fully integrated steel corporation with coke ovens, blast furnaces, steelmaking capacity, and finishing facilities. The background of the merger can be traced to the mid-1950s when Acme tried to reduce its dependence on outside sources for steel, for up to that time it was essentially a semiintegrated company which produced strip from slabs that were purchased from integrated steel companies.

In September 1956 Acme acquired the Newport Steel Company of Newport, Kentucky which had a capacity in seven open hearth furnaces and three electric arc furnaces to produce 600,000 tons of ingots per year. Part of this tonnage was to be used to ensure Acme of a source of extra steel when needed. This acquisition was considered an interim step, for in the 1956 *Annual Report* Acme announced that it had entered into a program to construct steelmaking facilities at its Riverdale plant. These were to consist of two basic oxygen converters with an annual capacity of approximately 450,000 tons which were to be supported by two hot blast cupolas designed to feed the oxygen converters with a continuous supply of molten iron made from cold pig iron and scrap. This program was undertaken to reduce Acme's dependence on outside steel companies for the slabs needed to operate its rolling mills. It had been dependent for slabs on the United States Steel Corporation for many years. In the 1950s the requirement was approximately 500,000 tons per year, and it was made clear by United States Steel, because of the heavy demand for steel and the low profitability which resulted from the sale of semifinished steel, that Acme would not be sold additional tonnage. There was no intention to cut the amount of tonnage supplied to Acme, but any increase had been virtually ruled out. Thus in Acme's 1956 *Annual Report*, it was stated:

> The principal supplier has been unwilling to increase the maximum amount it must deliver under its contract, and as a result, in recent years of high demand the Company has endeavored to supplement this contract supply by purchasing semi-finished steel from whatever other sources have been available. Usually this could be done only by paying a premium over the regular market price, and there is also no assurance that a future supply will be available.

With the completion of two basic oxygen converters having an annual capacity of approximately 450,000 tons, Acme had its own source of steel. Nevertheless, the company had to look to the outside for pig iron and scrap to feed the oxygen converters. The major portion of its pig iron was derived from a single integrated steel company. Thus dependence on the outside was still substantial, although it was moved one step further back toward the raw materials. However, melting pig iron and scrap in a cupola to provide molten iron for the oxygen converter was never a totally satisfactory operation.

In 1964 when the opportunity arose to merge with the Interlake Iron Corporation, one of whose blast furnace facilities was located in South Chicago, Illinois close to the Acme plant at Riverdale, the step was taken. The two companies complemented each other and in no way competed with each other. Interlake manufactured and sold pig iron, coke and a group of ferro alloys, while Acme manufactured and sold light flat rolled steel products and a number of other items, such as flexible conduit, stitching machines and metal furniture.

Interlake had three plants: one at Erie, Pennsylvania, with one blast furnace; a second at Toledo, Ohio, with two blast furnaces; and a third at South Chicago, with two blast furnaces. The plant at South Chicago had a capacity to produce almost 800,000 tons, and much of this was transported a short distance to the Acme plant at Riverdale Station in Chicago. The other two plants were engaged primarily in the production of iron for the foundry trade. Total annual capacity at the time of the merger was in the area of 1.8 million tons of pig iron. Acme, in addition to its Riverdale plant which contained two basic oxygen steelmaking vessels and a variety of rolling and finishing facilities, had the Newport plant at Wilder, Kentucky, which was referred to above.

The name of the new entity was Interlake Steel Corporation with headquarters in Chicago.

In the 1964 *Annual Report*, the first report of Interlake Steel Corporation, the organization was referred to quite accurately as "America's newest integrated steel producer." Interlake, in addition to its blast furnace facilities, had considerable holdings in ore and coal properties, not the least of which was a 10% interest in the Erie Mining Company, a large producer of iron pellets located in Minnesota. The merger was worked out by an exchange of stock whereby 7/10ths of one share of Interlake common stock was received for one share of Acme common stock.

Wheeling-Pittsburgh

The second merger in the basic steel industry during the 1960s took place between the Wheeling Steel Corporation and Pittsburgh Steel Company in 1968. The origins of this merger date back to April 1963, when Norton Simon, president of Hunt Foods, a West Coast based company, acquired between 9 and 10% of the stock in the Wheeling Steel Corporation, which was sufficient to obtain a place on its board of directors. In November 1964, less than two years later, Norton Simon at a special meeting of the board was elected chairman. Shortly thereafter, the composition of the board was changed as the number of members was increased from fifteen to eighteen. A number of the directors resigned and the places of these, as well as the three added, were filled by company officers and by some outside appointees of Norton Simon. Seven of the eighteen members represented the Simon interests. During 1965 there were other drastic changes in the top management of Wheeling Steel, as the board of directors was reduced from eighteen to seven, and a new president, Robert M. Morris, formerly associated with Monsanto Chemical Company, was brought into the management. The seven-member board of directors included no one with any extensive background in the steel industry.

When Simon took over in 1964 he was, according to the *New York Times*, harshly critical of the previous management and claimed it would take six or seven years to rehabilitate the company and that during 1965 the company might lose between $4 million and $8 million. The projection proved quite accurate, as the performance under the new management in 1965 resulted in a lose of $4.6 million and in 1966 in a loss of $7.8 million. Wheeling Steel was in deep financial trouble. The *1966 Annual Report* stated:

> Earnings in 1966 were penalized by the continuing excessively high operating costs which have plagued this company for years, the higher debt burden, higher depreciation charges, plus some $14.1 million of unusual costs resulting from heavy start-up expenses of the 80" hot strip mill and related facilities.

As a consequence of its operational difficulties, Wheeling was forced to purchase semifinished steel and also have some of its steel processed by other companies. At the same time, it suffered a decline in the quality of its output.

The *1966 Annual Report* was an interesting mixture of disappointment with the earnings as well as an attempt at optimism. Net sales for that year were at an all time high of $293 million on 1,665,000 tons of steel products shipped. Nevertheless, there was a net loss of $7.8 million.

A pricing policy which was to some extent nontraditional in the steel industry—namely, that the price at the time of order (PTO-pricing) would be charged the customer rather than the price at the time of delivery—was instituted in November of 1965. Also, the intensity and number of marketing and advertising programs was stepped up.

The production phase of the company's activities was entitled Operation Bootstrap in its *Annual Report* with the subcaption "When capital is short—stretch it." Here the report spoke of the installation of a new slabbing mill which would be large enough to give Wheeling Steel a slab capable of realizing the full potential of the new 80-inch hot strip mill. This, it was stated, would be in operation in late 1968 or early 1969. There was also to be a new 80-inch cold reduction mill. These facilities were to cost a total of $150 million when completed.

The facility modernization and expansion plan undertaken in the 1963 to 1965 period was completed by 1966. It included the installation of two basic oxygen converters, which replaced obsolete open hearths, a new 80-inch wide hot strip mill, which replaced the obsolete 66-inch wide mill, and a new 60-inch wide galvanizing line. These three facilities necessitated heavy outside financing, and some $140 million had been borrowed from insurance companies and banks.

As a result of the 1966 losses and the rather dim future prospects, the lending institutions began to urge management changes. Prior to this, Norton Simon resigned as chairman of the board and a director on November 20, 1966. His place was taken by Gustave L. Levy, a partner in Goldman Sachs and Company, a New York investment banking house. One other member of the board resigned during this period so that in late 1966 there were but six members on the board. The appointment of Levy was hailed by Simon, as he said, "The nature of the most serious problem now confronting Wheeling has changed. The current financial needs and longer range financial requirements are areas to which Mr. Levy's talents are unusally well adapted. At the same time, there are areas in which the necessary vigor and directness of my past actions in cleaning up at Whelling Steel might be an impediment to future accomplishment."

In early 1967 the lending institutions were quite active in seeking management changes, and so it was that on April 18, 1967 Pittsburgh Steel Company entered the picture through the purchase of 77,350 shares of Wheeling Steel stock from Hunt Foods. An equal amount was purchased by the Kettering family of Dayton, Ohio. Thus the holdings of Hunt Foods were reduced from 254,706 shares to 100,000 shares.

Immediately after this purchase, Robert M. Morris resigned as president of Wheeling Steel, as did three of the five directors of the corporation. Allison R. Maxwell, Jr. and Donald C. Duvall, president and executive vice president, respectively, of Pittsburgh Steel Company, took over the chairmanship and presidency of Wheeling Steel Corporation. On October 3, 1967 one of the two remaining directors resigned, leaving but one of the previous administration on the board of directors. On December 28, 1967 Hunt Foods sold its remaining 100,000 shares to Pittsburgh Steel, whose holdings increased to 177,353 shares, representing 8.2% of the common stock.

In April 1968 the board of directors of both Pittsburgh and Wheeling set up merger committees with the purpose of negotiating a union between the two companies. On September 25, 1968, the boards of directors approved the merger. It was then presented to the stockholders of both companies in early December, and the actual union became effective on December 5, 1968. The board of directors, which was elected on the following day, consisted principally of Pittsburgh Steel officers with two representatives from Wheeling Steel.

As a consequence of the merger, Wheeling-Pittsburgh moved into ninth place among the steel companies of the country. It had a raw steelmaking capacity of 4.3 million tons. This was accounted for by two basic oxygen furnace shops, one at the Monessen plant and the other at Steubenville. The facilities of both companies complemented each other insofar as the new continuous 80-inch hot strip mill at Steubenville provided a facility to roll high quality sheets. This supplemented the 66-inch semifinished hot strip mill at the Allenport plant. In all the merger provided larger size, a broader range of products, improved customer service, and a more effective use of the sales force.

Cyclops-Detroit

The third merger which took place in late 1970 was between the Cyclops Corporation and the Detroit Steel Corporation. Cyclops had a number of steelmaking subsidiaries, among them were: The Empire-Reeves Steel Division which encompassed its principal steelmaking facilities; the Sawhill Tubular Division which operated pipe mills and had to purchase the skelp from which pipe was made from the outside; the Universal Cyclops Specialty Steel Division which produced a considerable amount of stainless steel at its two plants located at Bridgeville and Pittsburgh, Pennsylvania.

The requirements of the Sawhill Tubular Division were in the area of 160,000 tons per year and Cyclops sought to merge with a company that had sufficient strip mill capacity to provide this tonnage. An attempt was made to merge with Sharon Steel in 1968; however, the Justice Department intervened and Sharon was taken over by NVF Company. Subsequent to this, Cyclops approached NVF to purchase Sharon but was again rebuffed by a Justice Department ruling. This was based on the fact that Sharon also produced stainless steel and it was claimed that the combination of Sharon and Cyclops would reduce competition in specialty steels. Still in search of a source of skelp, Cyclops sought to acquire Detroit Steel Corporation which had a fully integrated plant at Portsmouth, Ohio, with approximately 1 million tons of steelmaking capacity and a 56-inch wide hot strip mill.

In order to achieve the desired merger, Cyclops purchased 761,794 shares of Detroit Steel's common stock on March 24, 1970. This stock was owned by American Export Industries and represented 19.3% of Detroit Steel's common stock. The purchase price was $12 million in cash plus a note for $5 million more. The management of Cyclops had every intention of issuing a tender offer for the remaining Detroit stock in order to acquire control of the company, however, the tender offer was not made. In its place Cyclops entered into an agreement on May 28, 1970 with Cleveland Cliffs Iron Company whereby Cleveland Cliffs would tender for the Detroit stock and, if successful, Cyclops would transfer its shares of the Detroit stock to Cleveland Cliffs in return for the assets of Detroit Steel Corporation.

The interest of Cleveland Cliffs was inspired by the fact that Detroit Steel Corporation owned 1,130,150 shares of Cleveland Cliffs' stock, with a market value on December 31, 1969 of $45,206,000. This represented a 29% ownership in the common stock of Cleveland Cliffs and was, in fact, the principal asset of the Detroit Steel Corporation, which in 1969 posted a $229,000 profit on sales of $109.5 million, a mere 2/10ths of 1% on sales. Thus the agreement would provide Cyclops with the facilities of Detroit Steel and would allow Cleveland Cliffs to repossess a large part of its stock.

Subsequent to the agreement, Cleveland Cliffs made a tender offer for Detroit Steel common stock at $16.50 per share. The offer was terminated on June 19, by which time Cleveland Cliffs had purchased 2,521,212 shares of Detroit common, or approximately 64.4% of the outstanding shares. This added to the Cyclops holdings gave the combine 83.7% of the outstanding shares of Detroit common. A temporary corporation was set up called Cliffs-St. Clair Corporation to which Cyclops transferred its 761,794 shares of Detroit common and ultimately received the physical assets of the Detroit Steel Corporation. The temporary entity was then dissolved and Cleveland Cliffs retained the shares.

Cyclops actually took possession of the Detroit Steel Corporation's plants in late 1970. No annual report was issued for the Detroit Steel Corporation in that year; however, its performance resulted in a considerable loss due particularly to the General Motors' strike.

As a result of the acquisition Cyclops obtained the following properties: a fully integrated plant at Portsmouth, Ohio with a capacity of approximately 1 million tons of raw steel. The plant included coke ovens, two blast furnaces, one of which was inoperative; five open hearth furnaces (four of which were operative) rated at 320 tons per heat; a 56-inch hot strip mill; a 3-stand cold mill and a rod mill. In addition it received the Detroit Steel Strip Division located in Detroit, Michigan and Hamden, Connecticut with facilities for reducing hot strip to cold strip, and the Tex-Tube Division located in Houston, Texas with 3 electric resistance weld mills for the production of pipe from 1 inch to 6-5/8 inches in diameter. There were also raw material deposits which included a 20% interest in the Mesaba Cliffs Mining Company. The company had to purchase all of its coal since it had no captive mines, and at full operations very large scrap purchases were necessary.

The market for Detroit Steel Corporation was somewhat diversified, with 35% of output going to the automotive industry, 21% to Steel Service Centers, 17% to appliances and office equipment, 10% to construction, 7% to machinery and industrial equipment and tools and the remaining 10% scattered among a variety of customers. General Motors was the largest single customer accounting for 13.6% of the 1969 net sales.

The profit performance of Detroit Steel in 1969 was extremely poor, as has been noted. Total earnings were $229,000 on a very large sales volume of $109.5 million. In 1970 the situation was far worse as the company registered a loss; however, no report was printed for that year since it was absorbed by Cyclops before the year was out. Much of this loss was due to the General Motors' strike which took place in the fall of 1970 and lasted for ten weeks. The loss of Detroit Steel in 1970 was reflected in diminished earnings of Cyclops. Nevertheless, Cyclops acquired a strip mill and 1 million tons of steelmaking capacity at Portsmouth, Ohio. To put this mill into efficient operating order required very heavy capital investment. This included $10 million to reline and repair the blast furnace, and at least an additional $30 million for the modernization and improvement of other facilities. The Portsmouth mill was able to supply a considerable amount of Saw Hill Tube's requirements and this was one of the basic reasons for the merger.

National-Granite City

One of the most dramatic changes in recent years in corporate organization in the steel industry took place in August 1971, when a merger was finalized between National Steel Corporation and Granite City Steel Company. This represented not only a change in corporate structure in the industry, but also a change in the attitude of the Justice Department which permitted this merger. Because it involved one of the larger steel companies and a medium-size steel company, it was more significant in terms of structural change than the previous take-overs of Jones & Laughlin and Youngstown Sheet and Tube by interests outside the steel industry.

National Steel was the fourth largest company in the United States before the merger. Subsequent to the merger, it moved ahead of Republic Steel Corporation into third place. In 1970 National Steel had sales of $1.2 billion and ranked fifth among the steel producers in the country, although in terms of raw steel production it actually ranked fourth, with 8,353,000 net tons, or 6.4% of the total industry output. In terms of shipments, it accounted for 6,238,000 net tons, or 6.9% of the industry's total. Granite City Steel Company ranked thirteenth among the domestic producers in 1970 with 1,517,000 net tons of raw steel production and 1,251,000 net tons of products shipped. In terms of sales volume, it ranked eighteenth with $198 million in revenues.

Properties and Facilities. National Steel Corporation controls extensive reserves of both iron ore and coal, either directly or through interests in other companies. In the United States and Canada, it has reserves of approximately 180 million gross tons of high grade iron ore and controls reserves of about 950 million gross tons of low grade ore which can be upgraded or beneficiated. The company also has substantial coal reserves of 500 million net tons in Kentucky, Pennsylvania, and West Virginia.

National operates two fully integrated steel plants: its Weirton plant at Weirton, West Virginia and the Great Lakes plant near Detroit, Michigan. Both of these fully integrated plants contain blast furnaces, steelmaking capabilities, and extensive rolling and finishing operations. The output is predominantly light flat rolled products, such as sheets and tinplate. The basic oxygen process for making steel is in operation at both plants, in addition to which the Great Lakes plant has electric furnaces.

A third plant located at Portage, Indiana on Lake Michigan, some 40 miles east of Chicago, is a finishing mill known as the Midwest division. This began operations in 1962 and consists of cold reduction facilities for sheets, as well as facilities to produce tinplate and galvanized sheets. It receives hot rolled coils from the Great Lakes plant near Detroit and processes them into finished products. Another major facility is a blast furnace plant located near Buffalo, New York which produces pig iron for the trade.

In addition to these facilities, National owns extensive tracts of land near Corpus Christi, Texas; San Francisco, California; and Camden, New Jersey, on the Delaware River. There are no plans at present to build facilities on these sites. In connection with its raw materials, National has had close ties with the Hanna Mining Company over a long period of years.

National has a network of ten steel service centers which serve in part as outlets for its products.

Granite City Steel Company is essentially a one plant operation, with facilities located at Granite City, Illinois across the Mississippi River from St. Louis, Missouri. The plant has two blast furnaces, modern oxygen steelmaking facilities, a new 80-inch wide continuous hot strip mill and facilities for cold reduction of sheets, as well as three galvanizing lines. The cold reduction mill, a four-stand four-high facility, was installed in 1946 and is relatively obsolete. In some respects this mill is noncompetitive with modern units installed within the last ten years. An example of this is had from the fact that Granite City was forced to drop out of the tinplate business since it could not maintain quality competitive with other steel companies and was not able to produce double reduced tinplate.

As far as the supply of iron ore is concerned, the company owns a 50% interest in a pellet plant at Pilot Knob, Missouri; the other 50% is owned by the Hanna Mining Company. In addition, substantial purchases of ore are made from the Hanna Mining Company. Metallurgical coal for coke is purchased entirely from outside interests.

Reasons for the Merger. When National Steel built its Midwest plant at Portage, Indiana, the decision was made to install a finishing facility leaving room to build a fully integrated plant sometime in the future. As stated above, hot rolled coils were provided from the Great Lakes plant. However, with the installation of additional cold reduction capacity in that plant, the company was too short on hot rolled coils to provide for the needs of the Midwest plant from Great Lakes. Thus, in a manner of speaking, its facilities were out of balance and hot rolled coils had to be bought from other producers in the industry. Purchases were made from several companies, including Granite City and Youngstown Sheet and Tube. In May 1970 a five-year contract was drawn up between National and Granite City under which Granite City was to supply a minimum of 50,000 net tons of hot rolled coils annually, with a possibility of extending this, should conditions require it, to a maximum of 100,000 tons. This was a particularly attractive arrangement for Granite City since its hot strip mill had a much larger capacity than its relatively obsolete cold reduction facilities could absorb. As a consequence, the new 80-inch wide continuous hot strip mill produced less than 1.5 million tons annually, which was much less than 50% of its capability. The contract allowed Granite City to use more of its mill's potential.

The sales contract between the two firms brought on merger thoughts and talks, and early in 1971 an agreement in principle was reached between the two companies.

The merger, pending the Justice Department's approval was to be put into effect on July 29, 1971. This date was extended for two weeks and the merger was finally consummated without opposition from the Justice Department in mid-August, 1971.

Competitive Impact of the Merger. National Steel's explanation of the competitive impact of the merger, which the Justice Department did not challenge, stated that as far as relevant geographic markets were concerned, National's market was the entire nation, while Granite City was limited to a 25-state market, outside of which its sales were negligible. Through a careful analysis of the products which both companies made and the areas in which they were marketed, it was demonstrated that there would be very little, if any, reduction in competition by the merger. These products included carbon plates, carbon hot rolled sheets and carbon galvanized sheets. One of National's principal products, namely tinplate, was not produced at Granite City.

The Terms of the Merger. Under the terms of the merger, Granite City became a wholly owned subsidiary of National Steel, and each share of Granite City common stock was converted into 0.4445 of a share of National common stock, while each share of Granite City preferred was converted into 2.2 shares of National common stock. The relative prices of these stocks from 1969 through June 1971 are given in Table 2-3.

On April 16, 1971—the business day immediately preceding the announcement of the merger—Granite City common stock closed at $16.625 per share, while National closed at $44.125 per share.[1]

The Proxy Statement issued in connection with the merger stated the reason for the action:

> It is believed that the proposed Merger will result in greater and more diversified earning power for stockholders of Granite City through operating economies, the integration of the facilities and organizational strengths of National and Granite City and diversification of resources of the two companies. Management of Granite City believes that the Merger will add financial strength to its operations, facilitating modernization, replacement or addition of plant and facilities. Both managements expect that the combination of Granite City's modern hot strip mill with National's new finishing facilities will have long range benefits for stockholders of both companies.[2]

Benefits of the Merger. As a consequence of the merger, National Steel was able to supply its own needs for hot rolled coils without resorting to outside purchases, or the construction of an additional hot strip mill which would have entailed a capital expenditure of approximately $150 million. If such a mill were

Table 2-3
Comparative Market Prices, 1969, 1970, 1971

	National Common Stock		Granite City Common Stock	
	High	Low	High	Low
1969				
First Quarter	52 5/8	45 3/4	25 3/8	20 1/8
Second Quarter	56 7/8	45 3/4	22 1/4	18
Third Quarter	50 1/2	42 1/8	18 7/8	14 1/8
Fourth Quarter	48 3/4	40 1/2	15 7/8	11 1/4
1970				
First Quarter	42 1/2	37 1/8	16 3/8	11 5/8
Second Quarter	44 3/4	32 1/2	15 1/4	10
Third Quarter	40 1/2	32 1/2	12 1/2	10 1/8
Fourth Quarter	43 1/8	36 7/8	14	10 1/2
1971				
First Quarter	46 5/8	39 3/4	14 7/8	11 3/8
Period from April 1 through June 9, 1971	47	41	17 1/2	13 1/4

Source: Granite City Steel Company, *Proxy Statement for Special Meeting of Stockholders*, July 1971, p. 6.

installed at the Midwest division, it would have required additional steelmaking facilities in order to provide the slabs, since these were not available at any other National Steel plant. If the steelmaking facilities were basic oxygen converters they, in turn, would have had to be supported with additional blast furnace and coke oven capacity.

The total investment for the coke ovens, blast furnaces, basic oxygen converters, and a continuous caster to convert the raw steel into slabs would have been in the neighborhood of $250 million, so that the entire outlay needed to bring National into balance would have been approximately $400 million. It is doubtful that a company even as financially strong as National would have made such an investment at this time. Further, this would have added one more hot strip mill to the ten built during the 1960s, and thus increased the overcapacity that the industry now has for the production of hot rolled sheets.

The addition of a new mill by National would have been a most unwise decision, particularly since it would have to be a large mill that most probably would be underutilized. All of this was avoided by the merger.

From Granite City's point of view, the merger had definite advantages. Granite City would supply the hot strip deficit for National Steel and thus its mill, which was definitely underutilized, could operate at a higher percentage of its capacity with the attendant reduction in unit costs of the hot rolled coils. In addition, because of the obsolete condition of Granite City's cold reduction facilities, it would have been necessary for the company to invest substantial sums in new facilities in this segment of the plant, as well as in a continuous casting unit which would make slabs of an adequate size for the hot strip mill. Further, considerable investment was required for the rehabilitation of the blast furnaces and coke ovens. The capital outlay for all of these projects could have been in the area of $250 million, and with the debt structure that Granite City had (approximately 52% of its total capitalization is in long-term debt) it would seem that the possibility of raising the additional funds would have been quite remote.

As an alternative to this modernization program, the company could have temporarily renovated its facilities. However, this too would have been costly and would have provided a stopgap rather than a solution to the problem. Thus, without the merger and the benefits which accrued from it, Granite City faced a rather difficult future. As a consequence of the merger, its problems can be readily solved. Further, if National had installed its own capacity, Granite City would have lost one of its largest customers.

The merger was an extremely good fit which benefited both companies and spared the industry additional hot strip mill capacity.

Solutions to Investment Problems

In contrast to the National-Granite City merger, whereby capital cost benefits have been derived from a change in corporate structure, the denial of the attempted merger between Bethlehem and Youngstown Sheet and Tube demonstrates the relationship that existed in the 1950s and 1960s between duplicated steel investment and the government's opposition to steel company mergers. As a result of the court's decision, Bethlehem Steel moved into the Chicago market with a $1.3 billion investment in a new plant at Burns Harbor, Indiana. If Bethlehem had been allowed to merge with Youngstown, this investment would not have been necessary as the company would have modernized and expanded the Youngstown Sheet and Tube facility at Indiana Harbor, most probably at a cost of $500 million. In the past few years Youngstown Sheet and Tube has invested over $400 million in its Indiana Harbor plant, and if this $400 million is added to the $1.3 billion Bethlehem has invested at Burns Harbor, the total investment in the two plants amounts to $1.7 billion. Conservatively speaking, this means that an investment of at least $1.2 billion could have been avoided by the merger.

There is a touch of irony in the Bethlehem-Youngstown attempt to merge. The two companies had made a similar attempt in 1930, but it was denied by a decision of Judge Jenkins given on the 29th of December 1930 in Youngstown, Ohio. An appeal was made against Judge Jenkins' decision, and in August 1931, the Court of Appeals voted in favor of the merger. At that time, however, the Depression had engulfed the economy, and the value of Youngstown Sheet and Tube's stock, which stood at $140 a share in March 1930 as against $103 a share for Bethlehem on the same date, had dropped to $22 per share while Bethlehem's stock dropped to $27 per share. The *Annual Report* of Bethlehem Steel for 1931 stated:

> Due to changed conditions, it was deemed impracticable to carry out the agreement covering the proposed acquisition of the properties and assets of The Youngstown Sheet and Tube Company, referred to in our last previous report, and your Corporation in October, 1931, exercised its option under the agreement and cancelled it.

Because of the staggering amount of capital needed for a fully integrated, multiproduct greenfield site plant such as Burns Harbor, it is extremely doubtful that any single company in the steel industry will build this type of facility in the foreseeable future. There may be greenfield site plants built, such as that constructed by the United States Steel Corporation at Baytown, near Houston, Texas, but this plant is not comparable to that at Burns Harbor. It consists of two electric furnaces, a continuous casting unit and a plate mill, but does not have blast furnaces, coke ovens, or the diversified finishing facilities of the Burns Harbor plant. This type of plant has kept investment requirements, although substantial, to a reasonable minimum. However, the Baytown plant's output is limited to plates; consequently, it does not offer the diversification of product available at Burns Harbor, where both plates and various types of sheets are produced.

The high cost of facilities in the steel industry raises a serious question about future steel investment. The United States economy will be able to support 190 million tons of raw steel capacity by 1980. If this is to be provided, the industry faces the prospect in the next few years of having to expand its current capacity of 160 million to 165 million tons, while at the same time it must replace facilities that will become obsolete between now and the end of the decade.

There is one possible solution to the problem of providing enough steel capacity to care for the needs of the economy in 1980. This would involve joint ventures between existing steel companies to establish either a complete greenfield site, integrated multiproduct plant with all the latest technology, or a portion of such a plant, which would produce at least semifinished forms of steel. Such an installation could involve coke ovens, blast furnaces and basic oxygen steelmaking capacity, or large electric furnaces.

Joint Ventures

The concept of joint ventures should be thoroughly investigated because the prohibitive cost of a multiproduct, fully integrated greenfield plant makes it necessary that new capacity at new locations be undertaken by joint ventures involving two or more steel companies. Such plants, based on the newest steel technology, could provide low cost steel and would represent either additions to capacity or replacements of existing capacity at new locations. The latter is a particularly significant problem because many of our obsolete plants are located in areas of surplus steel supply, and if they are to be replaced, they should be located elsewhere in the United States.

The joint venture in the steel industry in the United States has been confined primarily to the development of iron ore deposits. In fact, in the past few years virtually every major iron ore deposit, outside of those developed by the United States Steel Corporation, has been a joint venture project. To mention but a few: there is the Erie Mining Company operating a pellet plant in Minnesota for the production of pellets from taconite ore and owned jointly by Bethlehem Steel with 45%, Youngstown Sheet and Tube with 35%, Interlake Steel Corporation with 10%, and the Steel Company of Canada with 10%; another major taconite pellet plant in Minnesota is operated by the Reserve Mining Corporation and is owned jointly on an equal basis by Armco Steel and Republic Steel; the Iron Ore Company of Canada with its deposits in Labrador and Quebec has as many as seven participants, while the Wabush Iron operation, likewise located in Canada, has several more.

One reason for the joint ventures is the tremendous investment involved in iron ore development as well as pellet plants for the beneficiation of low grade ore; a large plant requires over $300 million. It certainly seems that it is time the industry moved from joint ventures in ore to joint ventures in steel mill operations. No longer can the steel companies afford to duplicate expensive facilities. Had joint ventures been allowed earlier, the demands of the economy for sheet products in the 1960s and, for that matter, well into the 1970s could have been satisfied with six continuous hot strip mills instead of the ten which were installed in the past decade, ranging in cost from $80 to $150 million each. Given the proper conditions and the availability of steel, each of these mills could produce from 4 million to 4.5 million tons of hot rolled sheets annually. At the present time none of the ten is producing anywhere near this amount, and there is an eleventh mill currently being installed by the Steel Division of the Ford Motor Company.

There may be some market problems in a joint venture which involves finished products, however, joint ventures for the production of steel slabs and other semifinished steel certainly involve no market conflict. Two or three companies could band together to build a mill with coke ovens, blast furnaces, steelmaking facilities, and either continuous casting or a traditional slabbing and blooming mill to produce semifinished steel which could then be distributed to the participants to roll on their individual facilities. This would be a far more rational approach to the development of our steel industry than we have had in the past. It is an approach that is taken by many of the European companies and would not only eliminate duplication of costly investment but would also ensure a more complete use of the facilities involved. In one respect, it makes little sense to have two or three facilities that are operating at much less than their capacity if two or three companies could have a single facility operating at full capacity with the resultant reduction in cost per ton of steel produced. Such an approach would probably require approval by the Justice Department on a case-by-case basis to make sure that competition would be maintained and that consumers would not be disadvantaged.

There would be definite economic advantages in building a fully integrated plant producing at least slabs, if not some finished products, in the Midsouth, near New Orleans, Louisiana. Such a plant could easily be furnished with iron ore from overseas, could receive its coal by barge through the Ohio and Mississippi Rivers and would be in a position to export steel as well as to move it back up the Mississippi and other rivers for a maximum distribution throughout the central part of the United States. It could also ship by sea to the East and West Coasts. If electric furnaces were used at the plant and there was a need to furnish them with prereduced pellets, this location would be particularly favorable for the production of prereduced pellets since gas which is used in the operation would be more readily available here than in many other sections of the United States. This plant could represent an addition to capacity or the replacement of existing capacity which is now obsolete and located in areas that have an overproduction of steel.

The concept of a joint venture in steel making has been examined by Kaiser Steel and Armco Steel who agreed to study the possibility of such an operation in Australia. They have subsequently been joined by August Thyssen, a German steel company. Other U.S. companies have interests in steel operations abroad, albeit they are small as in the case of the United States Steel Corporation which has an interest in Altos Hornos in Spain. It is necessary that we think in terms of joint ventures for the United States because in the coming decade the steel industry is faced with new and critical problems. In five or six years demand will be great and funds to build capacity will be limited.

In the late 1940s there was a controversy over steel capacity. The Truman administration maintained that the industry was not producing enough to satisfy the demands of the economy, and, in fact, in his State of the Union address in 1949 President Truman threatened to install 10 million tons of steelmaking capacity to make up for the lack of production in the industry. At the present time many may feel there is adequate capacity here and should we run short in the future there will be millions of tons available from other countries to supplement our supply. This is indeed true for the next two or three years, but after that time the country may well be short of steel.

It is interesting to note that many of the foreign steel producers are cutting back on their plans to install additional steelmaking capacity. This is particularly true in Japan where revisions have been made of the projected investment in steel during the 1970s. It is therefore quite possible with the worldwide growth in demand that most steel planners expect in this decade that one billion net tons of raw steel will be necessary by 1980 to satisfy the world economy's demand. At the present time the world has a capacity for about 700 million net tons of steel. If the foreign steelmakers do not increase capacity as they had planned two or three years ago, and we do not make the needed addition to steelmaking facilities in this country, there could well be a worldwide shortage of steel by 1977 or 1978.

It should be stressed vigorously that at the present time there is plenty of steel capacity and that imports present a formidable problem. This, in part, has been responsible for cutting down the installation of additional steelmaking facilities. The companies have not been profitable enough to warrant the large outlays necessary, but the time will come in the not too distant future, if profits remain as they are or do not improve considerably, when the country will be wanting for steel and it may not be available anywhere in the world in large enough quantities to supply total demand.

As a result of the poor profit performance in the steel industry during the past few years a number of companies have found themselves in financial difficulty, to the extent that the very survival of some could be brought into question. Certainly these companies would not be able to install expensive facilities without help from others, and should such help be forthcoming, it will still represent a substantial investment to all the companies involved. There is a pressing need to investigate the possibility of more mergers to save some of the marginal steel companies. These mergers will be of little benefit if they are between the smaller ailing entities. They can only be of significance and help if they involve the absorption of a smaller company by a larger, more financially stable company. Such mergers, given a thorough examination in individual cases to make sure there is no restriction of competition, would leave us with a steel industry comprised of fewer but stronger units more able to compete among themselves and with foreign steelmakers. Like joint ventures, such mergers would decrease the likelihood of duplicated facilities and reduce the capital demands on individual companies.

New Steel Technology and Cost Reduction

In addition to structural change through joint ventures and mergers, a partial solution to the steel industry's investment problems can also be found through the use of new and emerging steel technology with a potential for reduced investment and operating costs. One consideration affecting capital spending prospects in the industry during the remainder of the 1970s is the variety of technology designed to revolutionize precisely those phases of the production process most readily adaptable to joint venture investments, namely, ironmaking, steelmaking, and the production of semifinished steel shapes.

Direct Reduction

One of the processes which hold promise for lower investment and operating costs in the steel industry is the direct reduction of iron ore, which produces a briquette or pellet containing better than 90% iron. This product can be used in electric furnace steelmaking to replace or supplement scrap, and, consequently, direct reduction provides a means of bypassing the blast furnace and also eliminates the need for coke ovens which are essential to blast furnace operations.

During the past two decades a number of processes have been developed to reduce iron ore directly to make it usable in steelmaking operations. Much of the impetus for this development came from those areas of the world where materials for economic blast furnace operations were not available. These were usually nonindustrialized countries with iron ore deposits and little or no metallurgical coal for coke.

There was relatively little interest and much less urgency to develop direct reduction in the industrialized steelmaking nations, such as the United States, Japan, and many of the European countries. Their steel industries were for the most part committed to the blast furnace, and if metallurgical coal for coke was not available, or had run low it was imported.

In recent years circumstances have changed, and much more interest has been shown in direct reduction by large steel producers. Consequently, more effort and capital have been applied to solve the problems involved, and results have improved. The circumstances which prompted this upsurge in interest have been:

1. The great desire to build a steel industry in the developing countries, many of which have ore in one form or another and little or no metallurgical coal for coke
2. The supply of metallurgical coal on a worldwide basis, particularly low volatile coal, is limited
3. Coke ovens have presented difficult air pollution problems
4. The capital investment required for a blast furnace and the accompanying coke ovens has grown substantially, and there is increased interest in the electric furnace which can use prereduced pellets

The last three reasons have compelled many of the steel producers in the industrialized countries to investigate and invest in direct reduction processes with the hope of supplying prereduced pellets or briquettes for the electric furnace. If this materializes, in place of a very expensive complex consisting of coke ovens, blast furnaces, and basic oxygen converters, steel can be economically produced in an electric furnace fed continuously by prereduced pellets manufactured in any one of a half dozen ways with a substantially lower capital investment. A large number of processes have been developed in the post-World War II period with varying success. To accomplish their objective, kilns, electric furnaces, low shaft furnaces, and other devices employing gas or coal as a reducing agent have been used. Some of these methods have been abandoned, others are in a precarious state, while some have been successful.

At the present time in the United States there are two plants installed and functioning for the production of prereduced pellets. Both of these were built by the Midrex Division of Midland-Ross Corporation. One is located on the West Coast at Portland, Oregon and furnishes pellets for Oregon Steel, and the other is located at Georgetown, South Carolina and furnishes pellets for the Georgetown Steel Company. The capacity of these units is limited; each is capable of approximately 400,000 tons per year. Midland-Ross has recently put another plant in operation at Hamburg, Germany, and also has plans to build a larger unit in the New Orleans, Louisiana area. When completed, this unit could furnish 1.2 million tons of pellets annually.

Another process which has been successful in operation is the HYL Process at the plant of Hojalata y Lamina Steel Company in Monterrey, Mexico. Several million tons of pellets have been produced by this process and used successfully in electric furnaces for a number of years.

In addition, the Stelco-Lurgi Process (SLRN), the only one which uses coal as a reducing agent, has been in operation in Canada on a pilot plant basis, while other operating units have been installed in various locations throughout the world.

Armco Steel has a facility for direct reduction under construction at Houston, Texas, and the United States Steel Corporation has completed one in Venezuela.

In must be stressed that the production of prereduced pellets represents a very small tonnage in relation to electric furnace requirements. It may well be 1975 or later before the various processes have been perfected and sufficient capacity installed to supplement or replace scrap as a feed material for electric furnaces on a large scale.

In many instances decisions will not be made to expand capacity for at least two or three years, so there is yet time to observe the progress made in the direct reduction process. It is quite possible if the process lives up to expectations that a combination direct reduction plant and electric furnace unit could be installed for considerably less capital investment than the conventional coke oven-blast furnace-basic oxygen furnace complex. Some estimates have put the direct reduction-electric furnace combination for equal tonnage at 40% less than the coke oven-blast furnace-basic oxygen furnace complex. This is at best an estimate, since no large-scale direct reduction process plant has yet been installed in the United States. It is, however, an alternative that should be given thorough consideration by any company planning to install additional steelmaking capacity in the next few years.

Steelmaking Technology

During the past twenty-five years a thorough revolution in steelmaking technology has occurred. With current circumstances and recent technological developments, this phase of steel operations is expected to undergo continued and more diverse change in the years ahead, with the likelihood of reduced investment and operating costs.

At the close of World War II, three steelmaking processes were used in the United States: the open hearth, the Bessemer converter, and the electric furnace. Of the 66.6 million net tons of raw steel produced in 1946, 91% was melted in open hearth furnaces, 5% in Bessemer converters and 4% in electric furnaces. Crucible steel production had been in a protracted decline, and by 1946 only 20 net tons of capacity remained in place, so that for all practical purposes the long-standing process had passed from the steelmaking scene. By 1971, when 120.4 million net tons of raw steel were produced, the share of total output accounted for by the open hearth, which exceeded 90% as late as 1957, had declined sharply to 30%; electric furnace production amounted to 17% of the total, and a major 53% share was accounted for by the basic oxygen furnace (BOF), first introduced in this country on a small scale in 1954. Production by the Bessemer process, which attained its postwar peak of 4.9 million net tons in 1951, declined rapidly thereafter and reached extinction in 1969.

During the remainder of the 1970s steel investment must provide for the continued introduction of new steelmaking technology as additional open hearth capacity becomes economically and mechanically obsolete. At present the industry is producing open hearth steel at the annual rate of approximately 35 million net tons, and many of the furnaces in use are not only obsolete but also represent a severe air pollution problem. In a number of instances, it would hardly be worth the investment to equip them with pollution control devices, and they will undoubtedly be scrapped. Consequently, the industry's current capacity of about 160 million to 165 million net tons will shrink unless these furnaces are replaced.

Basic Oxygen Furnace. Thus far the principal steelmaking method used to replace the open hearth has been the basic oxygen process. Most open hearth plants were constructed within integrated steelworks having blast furnaces and coke ovens, and the BOF with its hot metal requirements is geared to the continued operation of such ironmaking capacity, unlike the electric furnace which essentially has been a scrap-based process. The introduction of basic oxygen steelmaking was unquestionably one of the greatest technological breakthroughs in the steel industry during the twentieth century. The new method revolutionized the production of steel insofar as it cut the time required to melt a heat from an average of nine to ten hours in the nonoxygen open hearth to forty-five minutes. It also required less capital investment, and operating costs were reduced as well.

The actual steelmaking process in the basic oxygen furnace, which is a large pear shaped vessel opened at the top, begins when the furnace is tilted to accept the charges of scrap and hot metal. It is customary for the process to use approximately 28% scrap and 72% hot metal, but the percentages vary depending on the availability of these charging materials and the specifications required in the steel produced. A computer provides the operator with complete information on the weights of hot metal, scrap and flux, as well as the volume of oxygen needed to achieve given specifications in the end product. Once the charge has been completed, the furnace is turned upright and an oxygen lance is lowered from its gallery high above the furnace to a desired position relative to the charge, usually about six feet above it, after which the high purity oxygen is introduced. At this point, flux and other materials are added, and the process continues until the charge is refined to the desired grade and composition. The oxygen is usually blown for an average of twenty-two to twenty-four minutes. The molten steel is then tapped from the furnace, which is immediately prepared for the next charge.

Q-BOP Steelmaking. A recent technological advance, the Q-BOP, is actually a variation or refinement of the BOF and has attractive potential as an open hearth replacement. Basically, the process utilizes bottom-blown oxygen converters (in contrast to the BOF in which the oxygen is top blown through a suspended lance), and promises the advantages of lower capital investment and operating costs, as well as ready adaptability to existing open hearth shops. The investment cost advantages of Q-BOP relative to the BOF process must be considered from two aspects: first, as a replacement for open hearth capacity, Q-BOP, with the absence of the lance gallery and other overhead equipment common to the BOF, can be installed in existing open hearth buildings and can utilize existing cranes, resulting in a 40% to 50% capital cost saving; second, if Q-BOP is installed in a greenfield site plant, the saving in building steel required for the lance gallery and overhead equipment is estimated to cut investment costs by 5% to 10%.

One of the most significant operating advantages for the Q-BOP relative to the BOF is an improvement in hot metal yield from about 88% of charged metalics to between 90% and 92%. Other operating benefits claimed for the process are as follows: higher scrap utilization, up to 33% compared with 28% in the BOF, which is extremely helpful in plants short on hot metal capacity; higher oxygen efficiency and the use of supplemental fuels in the form of liquid or gaseous hydrocarbons; improved refractory life (by eliminating direct contact between the refractories and oxygen); reduced fume generation (by blowing the oxygen at lower velocities). The process nevertheless requires pollution control systems of nearly the same size and cost as the BOF.

The United States Steel Corporation is currently installing the Q-BOP at its Gary, Indiana and Fairfield, Alabama plants. The Gary installation is to consist of three 200-ton furnaces and is scheduled for a late 1972 start-up. At Fairfield two 200-ton Q-BOP vessels are being built as a replacement for open hearth capacity and are due to start operating late in 1973.

Electric Furnace. The capital costs involved in obtaining additional steelmaking capacity can be reduced significantly by the installation of electric furnaces in place of either of the oxygen converter processes discussed above. The electric furnace has a decided advantage over the oxygen converter in that it can use a 100% scrap charge and, consequently, does not require an expensive investment in blast furnaces and coke ovens, which are required to produce hot metal for the oxygen converter.

The postwar period witnessed a dramatic change in electric furnace operations as they shifted from small "specialty steel" facilities to large units producing predominately carbon steel. Electric furnace output increased from 2.6 million net tons in 1946 to 20.9 million net tons in 1971—an eightfold growth. Of 1971's output, 14.9 million tons were carbon steel, 4.7 million tons were alloy steel and 1.3 million tons were stainless steel. During the twenty-five-year period, over 150 electric furnaces with an inside shell diameter of eight feet or more were installed throughout the United States. Their capacities ranged from ten to 400 tons per heat; eighteen furnaces were capable of producing over 200 tons, while twenty-seven were rated at between 150 and 200 tons. The largest furnace, with a 32-foot shell diameter and 400-ton heat size, was built in 1970 at Northwestern Steel and Wire Company's Sterling, Illinois plant.

SIP Steelmaking. The submerged injection process (SIP) is designed to improve the performance of the open hearth furnace. It blows oxygen through tuyeres directly into the side and beneath the surface of the molten bath of hot metal, not through the furnace roof onto the top of the bath as is done in open hearths with oxygen lances. Sydney Steel Corporation, Sydney, Nova Scotia, has successfully produced 700 test heats of SIP steel and is now converting an existing open hearth shop to the new process, which consists of injection systems not only for oxygen but also for other fuels, natural gas or oil, as well as for powdered fluxing materials such as lime.

Based on limited experience, it is claimed that SIP results in a significant saving in the time required for melting a heat of steel; it is estimated to take approximately 90 minutes for an SIP open hearth to melt a 200-ton heat, much less than the five to six hours generally required in the best oxygen-lanced open hearths. Additional advantages claimed for the process include improved furnace and refractory life (particularly for refractories lining the furnace roof), increased quality control due to easy testing of the molten metal during the refining process, increased scrap melting with less effect on heat times than in oxygen converters, and added flexibility in the grades of steel that can be produced. It should be noted that the furnaces at Sydney Steel Corporation are tilting open hearths, of which there are very few in the United States. Some question has arisen about the difficulty of using the process in the stationary open hearth, which constitutes the majority of United States furnaces.

Continuous Casting

Continuous casting is one of the genuine technological breakthroughs achieved by the steel industry in the post-World War II period. The process bypasses several steps in the production of steel, as it eliminates pouring steel into ingot molds, stripping the molds from the ingots, placing the ingots into soaking pits to develop an even temperature and, finally, the primary rolling stage by which the ingot is rolled down into a semifinished form, that is, a slab, billet, or bloom. In the continuous casting process, the steel is tapped from the furnace into a ladle and then poured directly from the ladle into the continuous caster. It solidifies as it passes through and emerges as a slab, billet or bloom.

This process has been in operation on a commercial basis in the United States for approximately ten years. The first installation was at the Roanoke Electric Steel Corporation in Roanoke, Virginia in December 1962. Since then over 16 million tons of continuous casting capacity have been installed in large and small steel companies throughout the United States. For the first six or seven years, the installations were primarily small units producing billets ranging in size from 3 x 3 inches to 6 x 6 inches. Many of these were installed in the so-called mini-mills, installations of less than 300,000 tons annual capacity, which produce continuously cast billets from relatively small electric furnace heats and roll the resultant product into concrete reinforcing bars or small structural steel shapes. These plants, of which there are approximately 40 throughout the United States, were relatively successful operations from the start.

In the late 1960s continuous casting units for wide, thick slabs were installed, and there was considerable difficulty in making these units perform up to their rated capacity. At the present time, there are relatively few wide slab continuous casting machines in operation in the United States. These have been installed at Armco Steel Corporation, Inland Steel Company, Lukens Steel Company, McLouth Steel Corporation, National Steel Corporation, Phoenix Steel Corporation, Republic Steel Corporation and the United States Steel Corporation, which currently has two machines in operation. Until quite recently the wide slab casters have not performed up to their potential. However, in the past year many of the difficulties that have beset the operation have been solved, and the potential capacity has been reached in most instances. Therefore, it is safe to say that the savings which were projected for continuous casting are still to be realized and await more and more units to be installed in steel companies throughout the country. In fact, it is quite doubtful that a conventional slab mill will be built in the future in the United States. Consequently, during the present decade, there is no question that the industry will install more and more continuous casting units and will reap the savings involved.

A continuous casting unit, which takes the steel directly from the furnace and produces a semifinished form, insures a yield from the furnace to the semifinished form, be it slab, billet or bloom, of 95% or more. This compares most favorably with the conventional process whereby ingots are formed and rolled down on a mill at a yield of approximately 85%. The difference is due basically to the fact that each ingot must be cropped, fore and aft, as it goes through the mill to form a slab, whereas the continuously cast slab is produced from the entire ladle and must be cropped only at the beginning and end; thus there are but two croppings as compared to many in the conventional process.

Any fully integrated plant to be installed in the next decade, be it a joint venture among several steel companies or an individual effort, will most probably use continuous casting.

Investment for Pollution Control

The most pressing need for investment in the steel industry and, for that matter, in most other heavy industries is for equipment to control air and water pollution. Regulations by federal, state, and local governments have become increasingly stringent over the past few years, and these have been supported in many instances by the courts.

The steel industry has a number of old facilities which pose particular difficulty in the matter of air and water pollution; the principal ones among these are coke ovens and open hearth furnaces. New technology is being developed to solve the problems in question. However, as restrictions on the amount of pollutants allowed become tighter, it is more difficult to meet them. The open hearth problem in a number of instances has been aggravated by the use of oxygen. Fortunately for the industry, however, the open hearth furnace is being rapidly replaced by the basic oxygen steelmaking furnace and the electric furnace.

The new facilities are, for the most part, adequately equipped with pollution controls, and in many instances the costs of these controls represent as much as 8% or 9% of the total cost of the installation. The problem arises with the old facilities which are being phased out. Some open hearths have as much as four or five years of life, and adequate pollution control equipment would cost in excess of $1 million per furnace.

The coke ovens present an even greater problem, and in some instances there have been threats on the part of local authorities to shut down the ovens if some effective pollution control is not installed. In a recent case, one company estimated that it will spend in the next few years about $86 million on coke ovens in one plant to satisfy emission standards.

With the need to equip new facilities and existing facilities for both air and water pollution control, the steel industry must invest in the next four to five years something in excess of $2 billion. This investment will not produce any more steel in terms of quantity, it will not speed up the time required, nor will it result in improved quality. It is a social cost, and the urgency of the matter demands that this investment be made as soon as possible. As a result of this heavy commitment, less will be available in terms of capital funds for other needs. However, since the benefit is gained not by the individual company but by society, there should be some consideration given to assist the industry in providing for cleaner water and air. One possibility would be an investment tax credit on all pollution control equipment of 15% to 20%. The Revenue Act of 1969 permitted a company to write off antipollution equipment placed on existing facilities on a five-year basis; however, it made no such allowance for new facilities. Since the burden is so great and the benefit to the public so substantial, it would seem that tax consideration should be given to help finance antipollution equipment on both old and new facilities.

3 Voluntary Restraint Agreements on Steel Imports

In 1971, as in the prior six years, imports continued to be the primary problem facing the steel industry in the United States. The problem had its roots in the late 1950s as the steel industries of other countries finally recovered from the ravages of war and rebuilt their steel industries. Tonnages coming into this country increased during the 1960s and finally reached very large proportions in 1968. In that year, a record 18 million tons were imported. Table 3-1 gives the imports from 1957 through 1971.

It can be seen that the increase was substantial in 1959 when the steel industry here experienced a 116 day strike. After that time, there was a slight drop in imports for a year or two and then a gradual rise through 1964. In 1965—a year of labor negotiations in the steel industry—there was an abrupt increase of roughly 60% due to hedge buying against a strike. In the next two years this level was not only maintained but increased considerably so that in 1967 imports reached an all time high of 11.5 million tons. In 1968, another

Table 3-1
Imports of Iron and Steel Products, 1957-71 (Net Tons)

Year	Tonnage Shipped
1957	1,154,831
1958	1,707,130
1959	4,396,354
1960	3,358,752
1961	3,163,233
1962	4,100,039
1963	5,446,326
1964	6,439,635
1965	10,383,021
1966	10,753,022
1967	11,454,502
1968	17,959,886
1969	14,034,287
1970	13,364,474
1971	18,303,959

Source: American Iron and Steel Institute.

negotiating year, there was a startling increase of 60% as the figure reached virtually 18 million tons. In 1969 the figure dropped sharply to 14 million tons because the principal exporters to the United States placed a voluntary limit on their shipments here. The year 1970 witnessed a slight decline from the previous year, as steel demand in the European Community reached such proportions that the countries were not able to meet the quota tonnages in the United States. Another round of labor negotiations took place in 1971, and with it, despite the voluntary restraint agreements, imports reached an all time record of 18.3 million tons.

Origins of the Import Problem

After World War II, Japan and the industrialized countries of Europe had to rebuild their steel industries which were severely damaged. This was a process which took a number of years and was even more prolonged in the case of West Germany and Japan since limits were placed on their output of steel by the victorious allies. In 1947 steel output in West Germany was 3.3 million tons, while that of Japan was slightly in excess of 1 million tons. In the same year, the 84.9 million tons produced in the United States represented almost 57% of total world crude steel output.

During the 1950s, the picture began to change so that by 1955, although the industry in the United States established an all time high of 117 million tons of crude steel, this amounted to slightly less than 40% of the world's total. In the same year, the output in West Germany had increased to 27 million tons, while that of Japan, which had moved up more slowly, stood at 10.4 million tons. Other countries which had picked up appreciably were the United Kingdom, whose output was 22.3 million tons; France, which recorded 13.9 million tons; Belgium-Luxembourg, slightly in excess of 10 million tons; and Italy, which was just short of 6 million tons.

Between 1958 and 1962 there was a recession for the steel industry in the United States as production fell below 100 million tons in each of those years.

In 1960 with 99.3 million tons, the United States accounted for 26% of world steel production, while West Germany moved up rapidly to 37.6 million tons and Japan increased its output to 24.4 million tons.

It was clear that the industrialized countries of the world were making a remarkable come back and had achieved output figures far above their prewar highs. For example, the prewar record output of Germany was just short of 25 million net tons in 1938, and that of Japan was 8.3 million net tons in 1940.

By the early 1960s, the steel companies in Japan and Europe had expanded their capacity to a point where there was steel available for export, and the most lucrative market in the world was the United States. It was necessary for these companies to literally buy their way into the United States market because there was skepticism here about their ability to deliver and the quality of steel. Thus prices of $20 to $40 a ton less than the American prices were offered to consumers in the United States, which often represented a figure 30% below the United States price. To some extent they were successful, but at first only with such products as wire nails, barbed wire, wire rods, and concrete reinforcing bars. In 1958, for example, of a total of 1.7 million tons imported, concrete reinforcing bars were 473,000 tons, wire nails were 197,000 tons, and wire rods were 181,000 tons. In that year our exports were in excess of 2.8 million tons and were concentrated in such sophisticated products as tinplate, cold-rolled sheets, hot-rolled sheets and galvanized sheets, which together amounted to about 1.1 million tons.

Nineteen fifty-nine represented the turning point in the steel import-export picture. Because of the prolonged strike, the United States for the first time since the turn of the century became a net importer of steel as imports rose to 4.4 million tons and exports fell to somewhat less than 1.7 million tons. This was not received with particular concern because of the strike. However, it started a trend and from that time forward the United States has been a net importer of steel.

It is particularly noticeable that there has been a shift during the middle and late 1960s in the type of product imported. Barbed wire and wire nails have remained at virtually the same tonnage level that they represented in the late 1950s. However, as a percentage of the total they have declined drastically. The heavy tonnage items are now sheets, plates, and structural steel.

The advance in sheet imports is particularly notable and has resulted from the installation in Europe and Japan of modern hot strip mills and cold reduction mills for the production of both hot-rolled and cold-rolled sheets. It is interesting to trace the growth in cold-rolled sheet imports. In 1959 they were a mere 30,000 tons. The figure jumped to 70,000 tons in 1960. In 1961 they dropped precipitously to less than 5,000 tons. A recovery was made to 52,000 tons in 1962. In 1963 the figures for cold-rolled and hot-rolled sheets were combined and the total reached 527,000 tons, which compares to the combined total of hot and cold-rolled sheets for 1962 of 172,000 tons. This was the breakthrough, for in subsequent years imports of sheets advanced dramatically.

Table 3-2 gives the cold-rolled sheet import figures from 1964 to 1971.

The tonnages of cold-rolled sheets came principally from Japan and the European Community. In 1970 Japan accounted for 824,000 tons and the European Community for more than 1 million tons. The other principal suppliers were the United Kingdom and Canada. The sources were the same in 1971, but the tonnages shifted somewhat as Japan supplied a bit more than 1 million tons, while the European Community accounted for 1.85 million tons.

The principal reason why sheet imports—including cold-rolled, hot-rolled and galvanized—grew so rapidly was the installation abroad of large continuous hot and cold strip mills during the 1960s. In 1961 total imports of these three

Table 3-2
Imports of Cold-Rolled Sheets, 1964-71 (Net Tons)

Year	Tonnage
1964	409,063
1965	1,257,825
1966	1,169,578
1967	1,424,518
1968	2,918,228
1969	2,006,564
1970	2,291,171
1971	3,689,577

Source: American Iron and Steel Institute.

categories of sheets amounted to approximately 112,000 tons; by 1971 this figure had swelled to an amazing 7.8 million tons.

Both the European and Japanese companies were able to install these large units because the companies themselves had grown in size either through expansion or mergers with other companies. Prior to World War II, there were very few individual companies in Europe or Japan with more than a million tons of steel production. As a consequence, they could not justify the installation of large continuous strip mills and the accompanying cold-rolled facilities since they did not have the necessary tonnage of steel to keep these operating at something approaching capacity. This was also true in the early postwar period. However, in the late 1950s and the 1960s, the picture changed. In fact in 1970 the merger of the two largest companies in Japan, Yawata Steel and Fuji Iron and Steel, resulted in the largest steel company in the world, a position which the United States Steel Corporation had held since its organization in 1901.

The German steel companies, particularly August Thyssen, also expanded and acquired units by merger. This was true of the French, the British, and the Italians so that many of the European companies had attained large enough size to justify high tonnage equipment.

Import Penetration of United States Markets

Table 3-1 lists the imports from 1957 through 1971. The tonnages increased dramatically. It is also interesting to note the growth in market share which imports achieved over the same period. Table 3-3 lists net shipments by American producers, imports, exports, apparent consumption (which is the result of adding net shipments to imports and subtracting exports) and finally imports as a percentage of apparent consumption, or the American market.

Table 3-3
United States Foreign Trade in Steel Mill Products, 1957-71

Year	Net Shipments by U.S.A. Steel Producers	Imports	Exports	Apparent Consumption[a]	Imports as a % of Apparent Consumption
	(Thousands of net tons)				
1957	79,895	1,155	5,348	75,702	1.5
1958	59,914	1,707	2,823	58,798	2.9
1959	69,377	4,396	1,677	72,096	6.1
1960	71,149	3,359	2,977	71,531	4.7
1961	66,126	3,163	1,990	67,299	4.7
1962	70,552	4,100	2,013	72,639	5.6
1963	75,555	5,446	2,224	78,777	6.9
1964	84,945	6,440	3,442	87,943	7.3
1965	92,666	10,383	2,496	100,553	10.3
1966	89,995	10,753	1,724	99,024	10.9
1967	83,897	11,455	1,685	93,667	12.2
1968	91,856	17,960	2,170	107,646	16.7
1969	93,877	14,034	5,229	102,682	13.7
1970	90,798	13,346	7,053	97,109	13.8
1971	87,038	18,304	2,927	102,515	17.9

[a]Apparent consumption equals net shipments plus imports less exports.
Source: AISI and U.S. Department of Commerce.

The penetration rose from 1.5% in 1957 to 7.3% in 1964. In 1965 it jumped to 10.3% and in 1968 stood at 16.7%. As a result of the imposition of voluntary restraints on exports to the United States in 1969, it fell to 13.7%; in 1970 it remained about the same at 13.8%, but in 1971 it stood at 17.9%, a record high.

The market penetration was general throughout the United States, however, it registered a much higher percentage in the western and southwestern states than it did in the rest of the country. For example, in 1964 imports accounted for 17% of the southwestern market and 18.4% of the western market. By 1968 the percentage of the southwestern market had jumped to 32.7%, and that in the western market to 27%.

The southwestern market includes, Louisiana, Oklahoma, Texas and Kansas. The western market includes Montana, Idaho, Wyoming, Colorado, New Mexico, Arizona, Utah, Nevada, Washington, Oregon, California, Alaska, and Hawaii.

Despite the fact that the west and southwest had a fairly high percentage of penetration by imports, the area where most of the steel came in during the late 1960s was the Great Lakes and Canadian border. Of the 18 million tons that came in in 1968, 6.6 million tons came in through the Great Lakes and Canadian border. Three cities, Detroit, Chicago and Cleveland accounted for virtually all this tonnage. Detroit took in 2,941,352 tons, Chicago, 2,050,261 and Cleveland 1,142,000. The next largest tonnage came in at the East Coast ports which stretched from Boston to Miami, and also included Tampa, Florida. The total here was 4.7 million tons. New York City and Philadelphia accounted for 1.1 million tons and 1.250 million tons respectively. The Gulf Coast was next with 3.6 million tons. This was concentrated in the ports of New Orleans and Houston, which between them accounted for over 3 million tons. On the Pacific Coast, total imports were 2.6 million tons, with Los Angeles accounting for a little less than 1.6 million tons.

The offshore imports, which entered Puerto Rico, Hawaii, and Alaska, amounted to 375,975 tons, with San Juan accounting for 287,478 tons. Almost 50% of this tonnage was in concrete reinforcing bars.

In 1969 the total import tonnage for the United States declined, but the pattern of distribution remained about the same. Of the total of 14 million tons, 4.4 million came in at the Great Lakes and Canadian border; 3.5 million on the East Coast; 3 million on the Gulf Coast; and 2.6 million at the Pacific Coast ports. Offshore tonnage increased to slightly over 500,000 tons, with Puerto Rico accounting for about 306,000 tons of which 147,000 were concrete reinforcing bars.

Alaska showed a decided increase as the Japanese began to ship large diameter pipe for the trans-Alaska pipe line. In 1969 this amounted to approximately 80,000 tons. Hawaii accounted for 107,000 tons, 53,000 of which, or almost half, was tinplate, virtually all of which came from Japan.

The distribution of imports in 1970 followed the same pattern as in the previous two years, with 4.6 million tons coming in at the Great Lakes and Canadian border; 3.1 million tons on the Atlantic Coast; 2.6 million tons on the Gulf Coast and Mexican border; and 2.5 million tons on the West Coast.

Offshore tonnage increased to 637,000 tons, with Alaska as the leading importer with a total of 288,781 tons; 281, 216 tons were in large diameter pipe which came from Japan for the trans-Alaska pipe line.

The record of 18.3 million tons of imports which came in 1971 followed the same distribution pattern with one exception—total tonnages were greatly increased. The Canadian border and Great Lake ports accounted for virtually 7 million tons, while the East Coast witnessed the importation of 4 million tons. The Gulf Coast and Mexican border surpassed the East Coast by a slim margin as 4.1 million tons came in. Total imports on the West Coast amounted to 2.7 million tons, while the offshore tonnage fell off somewhat to 545,957 tons. This was due to a drop in the tonnage of pipe coming into Alaska from Japan. In that year, it fell to 90,172 tons as the Japanese completed the shipment of pipe for the trans-Alaska pipe line. Over the three years, the total tonnage of large diameter pipe shipped into Alaska amounted to approximately 450,000 tons.

The trans-Alaska pipe line, when and if it is constructed, will consist of 48-inch diameter pipe. This was supplied from Japan because there was no mill in the United States capable of producing pipe of this diameter. The limitation of the United States mills was 42 inches in diameter, with only one able to produce 44 inches. A number of American steelmakers claimed that they could have easily installed a mill which would have enabled them to produce this size pipe had they been given the opportunity to do so. It was estimated that such a mill could have been completed within a year.

The extent of imported tonnage in relation to the American market can be illustrated by a comparison of import tonnage with the shipments of the two leading corporations of the United States, namely, United States Steel Corporation and Bethlehem Steel. It can be seen from Table 3-4 that imports, in a manner of speaking, represented the third largest steel corporation in 1967. In 1968 they surpassed the tonnage of Bethlehem Steel and would have represented the second largest corporation. Because of the voluntary restraint agreements in 1969 and 1970, import tonnage fell to third place, but only by a slim margin. In 1971 when, for all practical purposes, the voluntary restraint agreements were scrapped, they rose to second place, 5.7 million tons ahead of Bethlehem and only 1 million tons behind United States Steel.

Table 3-4
Comparison of United States Steel Corporation and Bethlehem Steel Corporation Shipments with Import Tonnage, 1967-71 (Millions Net Tons)

Year	United States Steel Corporation Shipments	Bethlehem Steel Corporation Shipments	Import Tonnage
1967	19.8	13.1	11.5
1968	22.5	14.4	18.0
1969	22.4	14.5	14.0
1970	21.0	13.8	13.4
1971	19.3	12.6	18.3

Source: *United States Steel Corporation Annual Reports, Bethlehem Steel Corporation Annual Reports, American Iron and Steel Institute Statistical Reports.*

The increase in imports gave rise to calculation and speculation as to how many jobs in the United States could be created if imports were cut by a substantial amount. One source put it as follows:

> Growing imports have had an adverse effect on employment opportunities in the domestic industry, although this has been obscured somewhat by high rates of economic activity during the last few years. Many factors, of course, affect the level of steel employment, but a measure of the impact of imports can be found in the fact that it now takes about 6,000 employees in steel plants and offices and 1,200 employees in supporting activities to produce one million tons of steel mill products annually. On this basis, steel imports in 1968 meant the loss of

130,000 job opportunities, and the reduced level of imports in 1969 still represented over 100,000 lost job opportunities.[1]

Support for Steel Quota Legislation

It is not surprising that the mounting tide of steel imports gave rise to a definite movement for some type of protection against foreign steel producers. The first concrete move came in February 1966, when the American Iron and Steel Institute's Public Affairs Conference took place in Washington. A presentation of the difficulties faced by the industry was made by L.B. Worthington, chairman of the institute, in which he called for a temporary tariff to prevent what was termed unfair and unrestricted imports of steel products into the United States. The meeting was attended by many congressmen.

During the following two years, testimony was given before Congress by a number of steel officials aimed at controlling the level of imports. In 1967 the notion of a temporary tariff was abandoned in favor of a quota. Toward the end of that year, Senator Hartke offered a steel quota bill to establish orderly trade in iron and steel products. The bill proposed to limit steel imports to 9.6% of the United States market, which was the historical figure obtained in 1964 through 1966. Hearings continued and a substantial amount of support was mustered. In June 1968, the drive by the American steel industry was joined by the United Steelworkers of America and was stepped up in intensity. A most significant incident occurred on June 18, 1968 when Thomas F. Patton, chairman of the Republic Steel Corporation and I.W. Abel, president of the United Steelworkers of America sat side by side and testified in favor of the import quota bill.

Opposition to Steel Quota Legislation

The hearings before the House of Representatives Committee on Ways and Means and the Senate Committee on Finance brought considerable opposition to the quota bill. This came from two sources: the importers of steel, whose business would be affected by a quota bill and a number of industrialists and bankers whose companies carried on a large volume of international business.

Some industrialists and bankers expressed their opposition in a full page ad which appeared on November 17, 1967 in the midwestern edition of the *Wall Street Journal* and on November 20, 1967 in the eastern edition of the *Wall Street Journal.* The ad entitled "The $30,000,000,000 Question," expressed vigorous opposition to what were termed protectionist measures currently being discussed in Congress. It maintained that these measures, which included the steel quota bill, would gravely jeopardize American markets abroad, which amounted to $30 billion. It also stated that the quota bills would result in higher prices for the American consumer on a wide range of goods and services, and that quotas would violate the General Agreement on Tariffs and Trade which would entitle other countries, whose exports are affected, to retaliate against U.S. products. It said that this would be: *"retaliation to which we, as the world's largest exporter, are especially vulnerable."*

It went on to say:

> Foreign markets for American business, labor, and agriculture that now total some $30 billion—double the volume of only eight years ago—would be gravely jeopardized.

The businessmen and bankers considered the problem faced by steel and other industries, but this seemed to be outweighed by other considerations as they said:

> Mindful of the problems of some domestic industries faced with aggressive foreign competition, we must not lose sight of our still larger volume of foreign sales. *The game of international retaliation is simply one that we, as a nation, cannot win. . . .*
>
> *We fully recognize the very real problems and dislocations faced by some domestic industries—even industries with a high level of technical efficiency—in coping with low price imports. There are ways of meeting these problems without quantitative import restrictions.* One way we strongly support is the liberalization of the adjustment assistance provisions of the Trade Expansion Act, as the administration has proposed. We welcome the antidumping provisions of the Kennedy Round Agreements that should help maintain barriers against unfair competition. We particularly urge that the U.S. Government intensify efforts to reduce and eliminate existing non-tariff barriers abroad, and to provide more positive encouragement and support to our export effort.

It was signed by a group called "The Founding Members of the Emergency Committee for American Trade." The signees were:

William M. Allen
President
Boeing Company

George W. Ball
Chairman
Lehman Brothers International

William Blackie
Chairman
Caterpillar Tractor Company

R. Hal Dean
President
Ralston Purina Company

Henry Ford II
Chairman
Ford Motor Company

J. Peter Grace
President
W.R. Grace & Company

Patrick E. Haggerty
Chairman
Texas Instruments Inc.

H.J. Heinz II
Chairman
H.J. Heinz Company

William A. Hewitt
Chairman
Deere & Company

Robert S. Ingersoll
Chairman
Borg-Warner Corporation

James A. Linen
President
Time Incorporated

George S. Moore
Chairman
First National City Bank

David Packard
Chairman
Hewlett-Packard Company

Peter G. Peterson
President
Bell & Howell Company

Rudolph A. Peterson
President
Bank of America N.T. & S.A.

John J. Powers, Jr.
President
Chas. Pfizer & Co., Inc.

James M. Roche
Chairman of the Board
General Motors Corporation

David Rockefeller
President
The Chase Manhattan Bank N.A.

A. Thomas Taylor
Chairman
International Packers, Ltd.

Charles B. Thornton
Chairman
Litton Industries, Inc.

Arthur K. Watson
Chairman
IBM World Trade Corporation

Joseph C. Wilson
Chairman
Xerox Corporation

The 1968 Voluntary Restraint Agreements

Although there was also opposition to a steel quota bill by some forces within the administration of President Lyndon Johnson, support continued to grow in Congress, and under these circumstances, both the Germans and Japanese made a sudden and dramatic move in the summer of 1968. They proposed to both the Chairman of the Committee of Ways and Means of the House of Representatives and the Chairman of the Finance Committee of the Senate that they would place a voluntary restriction on their steel exports to the United States, and that they would make an earnest effort to induce other countries to join them. This proposal was based on the condition that hearings on the quota bill be stopped and that the bill be shelved. An agreement was reached in principle and the hearings were brought to a close.

No total tonnage figure was set for imports, nor was the rate of growth determined. Yoshihiro Inayama, president of Yawata Iron and Steel Company, Ltd., and chairman of the Japan Iron & Steel Exporters' Association, was quoted in the *American Metal Market* as saying he contemplated a 10% growth per year in imports. This was later denied. The figure of 7% was proposed by the Japanese as an annual growth factor. This caused much concern among American steelmen who could envision the growth in the American steel market being absorbed year after year by foreign producers.

Little was done to determine the tonnage figure for 1969 until November 1968. A number of discussions were held and finally the Europeans and Japanese settled on 14 million tons as the total import figure for the United States in 1969. In the course of the deliberations, the Europeans at first held out for 16 million tons, but the final compromise produced the 14 million ton figure.

The Japanese portion was 41%, or 5.75 million net tons and an equal amount was given to the six members of the European Coal and Steel Community which they were to apportion among themselves. This left 18% for the rest of the world. There was an attempt to include the United Kingdom in the arrangement, but it was unsuccessful. The arrangement further stipulated that the growth in these exports to the United States was to be limited to 5% in each of the two following years so that the total tonnage in imports in 1969 would be 14 million tons; in 1970 it would be 14.7 million tons and in 1971, 15.4 million tons.

In addition, both the Japanese and the Europeans stated their intention to maintain the product mix insofar as they were able. This portion of the statement was somewhat loosely worded, and indicated that the countries involved would try to maintain the product mix in approximately the same tonnage as before, or they would not change it substantially.

The letters containing the details of the voluntary restraint agreement were addressed to the secretary of state under the date of December 23, 1968:

To: The Honorable Secretary of State
Washington 25, D.C., U.S.A.

From: Yoshihiro Inayama, Chairman
Japan Iron & Steel Exporters' Association

SUBJECT: Statement of the Intention of the Japanese
Steel Industry

Statement of the Intention of
the Japanese Steel Industry

1. With the desire to assist in the maintenance of an orderly market for steel in the United States, the nine leading steel companies of Japan, namely, Yawata

Iron & Steel Co., Ltd., Fuji Iron & Steel Co., Ltd., Nippon Kokan Kabushiki Kaisha, Kawasaki Steel Corporation, Sumitomo Metal Industries, Ltd., Kobe Steel Works, Ltd., Nisshin Steel Co., Ltd., Osaka Iron & Steel Co., Ltd., and Nakayanma Steel Works, Ltd. gave assurances in their statement of July 5, 1968 that their steel mill product shipments from Japan to the United States would not exceed 5.5 million metric tons during Japanese fiscal year 1968. These nine companies account for approximately 85 percent of all Japanese steel mill products shipped to the United States. In the light of subsequent events and as a result of discussions concerning this matter with the representatives of the Government of the United States of America, they now want to make a new statement to the following effect.

2. With greater understanding of market conditions for steel in the United States, and with the cooperation of the medium and small steelmakers of Japan which account for the remaining 15 percent of shipments to the United States, the same nine leading steel companies wish to state their intention, subject to measures permitted by the laws and regulations of Japan, to limit the Japanese shipments of steel mill products to the United States to a total of 5,750,000 net tons during calendar year 1969.
3. During the subsequent two calendar years (through 1971), it is also their intention to confine the Japanese shipments within limits which would represent, at most, a 5 percent increase over 5,750,000 net tons in 1970 and over 6,037,500 net tons in 1971, depending upon demand in the United States market and the necessity to maintain orderly marketing therein. During this period the Japanese steel companies will try not to change greatly the product mix and pattern of distribution of trade as compared with the present.
4. This statement is made upon the assumptions: i) that the total shipments of steel mill products from all the steel exporting nations to the United States will not exceed approximately 14,000,000 net tons during 1969, 105 percent of 14,000,000 net tons in 1970, and 105 percent of 14,700,000 net tons in 1971, ii) that the United States will take no action, including increase of import duties, to restrict Japanese steel mill product exports to the United States, and iii) that the above action by the Japanese steel companies does not infringe upon any laws of the United States of America and that it conforms to international laws.

s/Yoshihiro Inayama
Chairman
Japan Iron & Steel Exporters' Association

The Honorable
Secretary of State
New State Building
Washington 25, D.C.
U.S.A.

Sir,

The associations of the steel producers of the ECSC united in the "Club des Siderurgistes," to wit:

- Associazione Industrie Siderurgiche Italiane ASSIDER, Milan represented by Prof. Dr. Ernesto Manuelli

- Chambre Syndicale de la Siderurgie Francaise, Paris represented by the President, Mr. Jacques Ferry
- Groupement des Hauts Fourneaux et Acieries Belges, Brussels represented by the President, Mr. Pierre van der Rest
- Groupement des Industries Siderurgiques Luxembourgeoises, represented by the President, Mr. Rene Schmit/Luxembourg
- Vereniging de Nederlandse Ijzer- en Staalproducerende Industria, represented by Mr. Evert van Veelen / Ijmuiden
- Wirtschaftsvereinigung Eisen- und Stahlindustrie, Dusseldorf represented by the President, Bergassessor Dr. Hans-Gunther Sohl

referring to the repeated talks they have had in this matter with representatives of the Government of the United States in behalf of the sustenance of liberal international trade in steel and to assist in the maintenance of an orderly market for steel in the United States declare the following:

1. It is their intention to limit the total ECSC deliveries of steel mill products, i.e. finished rolled steel products, semis, hot rolled strip, tubes, and drawn wire products, to the United States to 5.750.000 net tons during the calendar year 1969.
2. It is also their intention in the calendar years 1970 and 1971 to confine their deliveries within limits which would at the utmost represent for the year 1970 a five percent increase over 5.750.000 net tons and for the year 1971 a five percent increase over 6.037.500 net tons.

During the named periods the ECSC producers will try to maintain approximately the same product mix and pattern of distribution as at present.

This statement is based on the assumption

A) that the total shipments of steel mill products (finished rolled steel products, semis, hot rolled strip, tubes, and drawn wire products) from all the steel exporting nations to the USA will not exceed approximately 14 million net tons during 1969, and five percent over 14 million net tons in 1970, and five percent over 14.7 million net tons in 1971, and

B) that the United States will take no action to restrict ECSC steel mill product exports to the USA like

a) quota systems

b) increase of import duties

c) other restrictions on the import of steel mill products to the USA.

This proposal of the ECSC steel producers is made provided that it does not infringe on any laws of the United States and that it conforms to international laws.

/ s / - Ernesto Manuelli - / s / - Jacques Ferry -
/ s / - Pierre van der Rest - / s / - Rene Schmit -
/ s / - Evert van Veelen - / s / - Hans-Gunther Sohl -

The secretary of state transmitted them to the chairman of the House Committee on Ways and Means and the Senate Committee on Finance on January 14, 1969, with the following letter:

The Honorable
Wilbur D. Mills
Ways and Means Committee
House of Representatives

January 14, 1969

Dear Mr. Chairman:

The President has asked me to transmit to you communications received from the steel industry of Japan and the steel industries of the European Coal and Steel Community (ECSC) expressing the intentions of these industries to limit their exports of steel mill products to the United States in the years 1969 through 1971.

We estimate that as a result of the export limitation of the Japanese and ECSC producers, which together provide about 82 percent of our steel imports, total imports will amount to about 14 million net tons in 1969, about 14.7 million net tons in 1970 and about 15.4 million net tons in 1971. Other major foreign producers have not formally offered to cooperate in the voluntary export limitations but, as a practical matter, are expected to maintain their exports at levels which yield the estimates stated above.

Sincerely yours,

/ s / Dean Rusk

Two important exporters of steel to the United States, namely, Canada and the United Kingdom, did not sign the agreement and, therefore, were not obliged to follow its details. Likewise, all the other steel producing countries in the world were not included and thus technically were not under any restraints.

In 1969 the quota was adhered to very strictly as imports stood at 14,034,287 tons, a mere 0.2% more than the voluntary restraint level. The United Kingdom and Canada made substantial reductions in their exports to the United States in 1969 compared with the previous year. The Canadian figure dropped from 1,242,898 tons to 804,659 tons, while that of the United Kingdom declined from 1,302,299 to 893,513 tons.

The amount of steel received from Japan in 1969 was 6.2 million tons, or about one half a million above the quota. This was explained by the fact that a substantial tonnage was shipped from Japan in December 1968 and was not received in the United States until January 1969. Thus although it arrived in the quota year, it was sent the previous year.

The Europeans fell substantially below their quota with total shipments to the United States of 5.2 million tons in 1969.

Among the nonquota nations, a few increased their exports to the United States in 1969. In several cases the tonnage represented a substantial percentage increase, however, in absolute terms, it was relatively slight. For example, Mexico stepped up its exports to the United States from 162,608 tons in 1968 to 182,583 tons in 1969. India's exports rose from 13,090 tons to 30,378 tons in the same period, and the Republic of South Africa increased its tonnage sent here from 2,633 to 42,000. Small tonnage increases also came from the Korean Republic, Taiwan, and Czechoslovakia.

The sharpest decreases in 1969 were in the Japanese exports, which dropped by more than one million tons, and those of West Germany, which declined by approximately the same amount. Imports from Belgium-Luxembourg fell off by approximately one-half million tons in 1969.

In 1970 total imports, which amounted to 13,364,474 tons, fell below not only the 1969 level, but were actually 8.3% under the 14.7 million ton figure which was allowed in keeping with the voluntary restraint agreements. Substantial declines took place in Japanese tonnage, which dropped to 5,939,475, as well as that from the European Community, whose shipments to the United States fell to 4,572,615 tons. The drop in the European tonnage was due to a substantial boom which took place in Europe and absorbed much of the steel output in the European countries during the first six months of 1970.

In the last four months of the year, business slowed considerably, and strenuous efforts were made to increase exports. However, because of the time lag between the placement of an order and its shipment, which is normally two to three months, steel exports from Europe to the United States fell short of the quota tonnage.

A number of the nonquota countries increased their exports above the 1969 level. These included Canada, where the tonnage rose from 804,659 to 1,105,363, or an increase of 300,000 tons. Others which recorded increases somewhat significant on a percentage basis but limited on a total tonnage basis, were the Republic of South Africa, Spain, Argentina, and India. The last named country had a relatively substantial tonnage increase from 30,378 to 106,076. The United Kingdom, on the other hand, showed a decline as the figure dropped from 893,513 to 824,193.

It is an interesting fact that while imports fell 25.6% from 1968 to 1970, virtually the same revenues were generated by this reduced tonnage. In 1968 the cost of steel imports was $1.976 billion and in 1970, it was $1.967 billion. The foreign steelmakers were able to maintain dollar volume for two reasons: first, there was a substantial increase in the price level. In one case, wire rods, the price advanced over a period of approximately two and a half years from $80 to $130 per ton. Second, as tonnage diminished there was a shift to the higher priced items. This was particularly evident in the specialty steels, including stainless, where there was a decided shift from semifinished products to cold-rolled sheets. In 1968 total stainless steel imports were 174,031 net tons; in 1969, this figure rose to 182,224, and in 1970 fell back to 177,209. However, in the same three-year period cold-rolled stainless sheets moved from 69,012 tons in 1968 down to 62,739 tons in 1969, and then rose abruptly to 75,681 tons in 1970 to constitute 42.7% of all the stainless imports. Semifinished products over the same period moved from 39,602 tons of imports in 1968 to 49,044 tons in 1969, and then dropped sharply to 27,015 tons in 1970. It should be noted that the stainless steel figure did increase in tonnage in 1969, when most other products were cutback, and in 1970, although there was a decline, the total tonnage was still above the 1968 figure.

Tool steels witnessed a minor increase between 1968 and 1969 as they rose from 15,162 tons to 15,253 tons. In 1970, however, this tonnage increased substantially to 17,635. This category of steel products does not constitute much in the way of tonnage, but it is a very valuable product and often sells for several dollars per pound.

In the third and last year of the voluntary restraint agreements, 1971, there were some changes in the international picture which materially affected the imports of steel. To begin with, the State Department reopened negotiations for an extension of the voluntary restraint agreements. These negotiations proceeded in a somewhat leisurely fashion, and the outlines of an agreement began to take shape. The discussions involved an extension of the voluntary agreements for a second three-year period, with the annual increment to be cut from 5% to 2.5%. In addition, fabricated steel was to come under the quota. The Europeans were particularly anxious to exclude 48 inch diameter pipe from the quota. They maintained there was no mill in the United States capable of making such pipe and, therefore, they should not be restricted in their sale of this product here. The Americans countered with the statement that if a pipeline of such a large diameter were open for bids, such a mill would be built in the United States.

As the voluntary restraint agreement discussions proceeded, business in general in Europe and Japan continued to decline, and the steel companies in these countries strove to export as much as possible. As a consequence, the 15.4 million ton figure, which was to be the year's full allotment, was shipped into the United States by October 1971, at which time imports had reached 15.515 million tons. Technically speaking, it was maintained by the Europeans that the Americans had violated the quota agreement when the Nixon administration put a 10% surcharge on all imports into the United States as of August 15, 1971. They argued that the quota agreement entered into in late 1968 contained an understanding that during the three-year period no tariff or mandatory quota of any kind would be levied against steel. However, it must be observed that while this was technically correct, the lead time between the placement of an order and the receipt of the steel is usually three months. This indicates that the orders placed for delivery through October were on the books of foreign steel producers before the 10% surcharge was announced. In addition, for several weeks after the 15 of August, the import situation was in a state of turmoil and order taking to a great extent was suspended. All of this indicates that much of the tonnage which came into the United States during November had also been ordered before the imposition of the surcharge on August 15.

By the end of 1971, a record 18,303,959 tons of steel products had been imported. This was approximately 2.7 million tons above the voluntary restraint agreement level. The revenues paid out for this tonnage were $2.636 billion, as opposed to receipts from American exports of $576 million for 2,827,377 tons of steel products shipped abroad. The deficit on steel alone was over $2 billion. Clearly, the voluntary restraint agreement had been scrapped.

It has been maintained that much of the increase in 1971 came from nonquota countries. There is something to be said for this as Table 3-5 indicates. This table gives the imports received from nonquota countries for the years 1968 through 1971. Table 3-6 gives the imports from quota countries during the same period. Although it can be seen from Table 3-5 that on a percentage basis there was a substantial increase from nonquota countries, the actual increased tonnage was not nearly as great as that from the quota countries.

For example, Japan increased its shipments by 16.4%. On a percentage basis this was nowhere near as great as the increase from Poland, which was 160%. However, the 16.4% represented a million tons, whereas the 160% increase represented 142,000 tons. The same can be said of the Philippines, where shipments increased from 9268 tons to 91,915 tons, or about 1000%. However, the 1000% accounted for 82,000 tons as opposed to a 60% increase from Belgium-Luxembourg, which accounted for about 742,000 tons. The largest increase from a nonquota country came from the United Kingdom, where imports advanced by over one half million tons, or 65%. The largest percentage increase from a quota country came from Italy, where exports to the United

Table 3-5
Imports from Nonquota Countries, 1968-71 (Net Tons)

Country	1968	1969	1970	1971
Mexico	152,680	182,583	190,281	349,273
Poland	159,752	147,693	89,825	232,031
Spain	15,357	10,290	42,433	197,869
Argentina	177,870	99,501	118,554	147,302
Korean Republic	3,015	8,569	26,611	133,116
Sweden	83,982	80,332	73,448	79,018
The Philippines	–	46	9,268	91,915
Czechoslovakia	13,837	21,033	20,805	41,385
Brazil	165,580	37,894	19,456	71,793
Taiwan	11,606	24,627	33,845	44,564
United Kingdom	1,302,299	893,513	824,193	1,357,345
Canada	1,242,898	804,659	1,105,363	1,272,847
India	13,090	30,378	106,076	9,586
Austria	27,193	28,073	20,490	18.252
Republic of South Africa	2,633	42,000	74,486	17,416
Australia	154,428	102,940	66,002	46,156

Source: American Iron and Steel Institute, unpublished material.

Table 3-6
Shipments Received from Quota Countries, 1968-71 (Net Tons)

Country	1968	1969	1970	1971
Japan	7,294,380	6,252,654	5,935,475	6,908,372
ECSC				
Belgium-Luxembourg	2,124,213	1,673,888	1,240,679	1,983,000
France	1,202,696	1,018,896	934,045	1,564,000
West Germany	3,068,838	1,932,535	1,752,427	2,509,000
Netherlands	430,712	430,506	502,265	613,000
Italy	270,894	143,684	143,199	505,000

Source: American Iron and Steel Institute.

States rose from 143,199 in 1970 to 505,196 in 1971, an increase of approximately 350%.

Table 3-7 gives the tonnages by type of product, and it can readily be seen that the principal products imported were sheets, both hot-rolled and cold-rolled, as well as galvanized. Other products which registered more than a million tons were structurals, wire rods, plates, and welded pipe and tubing.

The 1972 Voluntary Restraint Agreements

The voluntary restraint agreement which was made in December 1968 was scheduled to terminate on December 31, 1971. There is some question about whether or not the agreement had been terminated by the 10% surcharge placed on imports by President Nixon on August 15, 1971. Judging from the amount of steel that was imported in 1971, it would be reasonable to say the agreement did terminate on August 15, 1971, or at least was ineffective after that date.

The negotiations to extend or renew the voluntary restraint agreement began about mid-1971, and an agreement was finally reached and signed on May 2, 1972. There were a number of reasons for the delay in reaching an accord. One of these was the insistence of the State Department that some definite limitations be placed on the amount of specialty steels to be imported. Another was the desire to include fabricated structural steel under the quotas. Further, the Europeans wanted to exempt large diameter pipe of 48 inches and over, since, as mentioned previously, there is no mill in the United States that can produce this size.

The Japanese reached an accord considerably before the Europeans. This was not surprising since the Europeans had seven nations involved and one of these, Italy, had increased its exports to the United States by 350% in 1971 and, consequently, wanted a larger share of the quota.

Table 3-7
Imports of Iron and Steel Products, 1971 (Net Tons)

Steel Mill Products	Shipments
Ingots, Blooms, Slabs and Billets, etc.	274,414
Wire Rods	1,544,359
Structural Shapes (Plain 3" and Over)	1,482,669
Sheet Piling	89,208
Plates	1,573,665
Rails and Track Accessories	38,247
Wheels and Axles	30,574
Concrete Reinforcing Bars	514,810
Bar Shapes Under 3"	550,211
Bars	
Hot Rolled, Carbon	807,564
Alloy	102,535
Cold Finished	117,636
Hollow Drill Steel	2,394
Pipe and Rubing	
Welded	1,406,620
Other	434,505
Round and Shaped Wire	539,380
Flat Wire	24,020
Bale Ties	600
Galvenized Wire Fencing	47,319
Wire Nails	293,586
Barbed Wire	58,142
Black Plate	7,454
Tinplate	416,155
Terne Plate	1,459
Sheets	
Hot Rolled	2,664,348
Cold Rolled	3,689,577
Coated (Inc. Galvanized)	1,409,921
Coated Alloy	90
Strip	
Hot Rolled	36,836
Cold Rolled	49,662
Hot and Cold-Rolled Alloy	28,646
Sheets	
Other, Electric Coated	67,353
Total	18,303,959

Source: *American Iron and Steel Institute Statistical Report for 1971*, p. 34.

As it was finally signed, the agreement differed substantially from the 1968 version. The United Kingdom was included among the European countries, and a number of stipulations on the tonnage of specialty steels were spelled out. The European version stated that total shipments in 1973 would be 1% higher than the 8,013,794 net tons to be shipped in 1972. In 1974, total exports from the European Community and the United Kingdom, which will be part of the community at that time, will not exceed the 1973 shipments by more than 2.5%. Thus 1973 shipments will be 8,093,573 and 1974 shipments will rise to 8,295,910.

The Japanese letter of agreement stated that they would not exceed 6,498,059 net tons in 1972, and this will increase by 2.5% for both 1973 and 1974. Consequently, the total tonnage for those two years will amount to 6,660,510 in 1973 and 6,827,023 in 1974.

Specialty steels, including stainless steel, were limited to definite tonnages. The Japanese specified that they would reduce shipments of stainless steel, tool steels, and other specialty steel to the following amounts:[2]

	1972	1973	1974
Stainless Steel	72,463	68,840	70,561
Tool Steel	893	915	938
Other Specialty Steel	155,935	141,901	145,448

This would seem to be a drastic reduction from the 1971 figures which for these categories were much higher. For example, stainless steel shipments from Japan in 1971, as recorded by the U.S. Department of Commerce, amounted to 115,901 tons. However, the Japanese figure for stainless shipments in 1971 was 99,412 tons. The discrepancy arises from the categories included under stainless steel by the Japanese and the U.S. Department of Commerce. The Japanese do not include wire and plate. Consequently, the cutback to the new figure of 72,463 was a reduction from 99,412 and not 115,901. This is nevertheless a substantial reduction of about 27%.

The Japanese accord also provided for a limitation on geographic distribution in the United States. They agreed that no more than one-third of their total shipments of steel mill products would go through ports on the Pacific Coast. The Japanese also agreed to limit fabricated stainless steel in the calendar year 1972 to an amount no more than 2.5% greater than 1970 with subsequent annual increases of 2.5%.

Both the European and the Japanese commitments stressed that cold finished bars would be confined during 1972 to a level of not more than 2.5% above 1970 with annual increments in growth after that time of 2.5%. There is a slight disagreement on this measure since the European letter simply refers to cold finished steel bars, while the Japanese refer to cold finished carbon steel bars.

It should be noted that the agreement limiting cold finished bars does not prevent a shift within the category from the lower to high-priced items, while the statement on specialty steels does call for a pledge not to depart significantly from the product mix developed in previous years.

The letters which spell out the agreement are given below:

May 2, 1972

Dear Mr. Secretary:

The Associations of the Steel Producers of the European Coal and Steel Community (ECSC) and the Association of Steel Producers of the United Kingdom, declare their intention to limit through December 31, 1974, exports of steel mill products to the United States. The following are the terms under which these exports will be regulated:

1. *Total Exports and Growth Rate*

The total quantity of steel mill products to be shipped to the United States by the above-mentioned producer associations will not exceed 7,270,000 metric tons (8,013,794 net tons) during calendar year 1972. The growth rate of exports of steel mill products will not exceed one percent (1%) in calendar year 1973 over the calendar year 1972 limit of shipments and the growth rate of exports of steel mill products will not exceed two and one-half percent (2.5%) in 1974 over the calendar year 1973 limit of shipments.

In accordance with this growth rate, such shipments will not exceed 7,342,000 metric tons (8,093,573 net tons) in calendar year 1973 and 7,525,550 metric tons (8,295,910 net tons) in calendar year 1974.

The above-mentioned producer associations will endeavor to ensure that shipments in either calendar semester of a given year will be held to no more than sixty percent (60%) of the total annual quantitative limit.

2. *Product Mix*

Within the foregoing limitations, the above-mentioned producer associations will limit their shipments of stainless, alloy and high speed steel mill products which are covered by the ECSC Treaty to total quantities not to exceed the following (metric tons):

Stainless Steel Mill Products	
1972	16,873
1973	15,871
1974	16,268
Alloy Steel Mill Products	
1972	77,927
1973	68,351
1974	70,059
High Speed Steel Mill Products	
1972	120
1973	99
1974	101

Within these total quantities the above-mentioned producer associations will not depart significantly from the product mix in each calendar year as compared with the pattern of such shipments to the United States over recent years.

Concerning high speed steel mill products and stainless and alloy steel mill products which are not covered by the ECSC Treaty, the above-mentioned producer associations will use their best efforts, consistent with legal requirements, to induce producers who are not members of the above-mentioned producer associations to limit their shipments to the following quantities (metric tons).

High Speed Steel Mill Products

1972	360
1973	370
1974	379

Stainless Steel Mill Products

1972	7,211
1973	6,633
1974	6,800

Alloy Steel Mill Products

1972	34,993
1973	35,034
1974	35,910

Exports of the aforementioned products will be regulated by reference to the classifications of the Brussels Tariff Nomenclature (BTN). A concordance of the BTN with the relevant classifications contained in tariff schedules concordance of the BTN with the relevant classifications contained in Tariff Schedules of the United States Annotated (TSUSA) is appended.

With respect to other categories of steel mill products, the above-mentioned producer associations will generally maintain the product mix in each calendar year in accordance with the average shipments for the period 1969, 1970, 1971.

With reference to exports of cold finished steel bars, special efforts will be made, consistent with legal requirements, to confine shipments during calendar year 1972 to a level of not more than two and one-half percent (2.5%) greater than during calendar year 1970 and to maintain a growth rate of two and one-half percent (2.5%) of the preceding year's limit during each of calendar years 1973 and 1974.

3. *Fabricated Structural Steels*

Such products are not considered steel mill products and are not within the limitations set forth in paragraph 1 above. However, with reference to shipments of fabricated structural steel products according to the definitions of the BTN, special efforts will be made, consistent with legal requirements, to confine the shipments by the aforementioned producer associations during calendar year 1972 to a level not more than two and one-half percent greater than during calendar year 1970, outside the quantitative limit as described in paragraph one above, and to maintain a growth rate of two and one-half percent (2.5%) of the preceding year's limit during each of calendar years 1973 and 1974.

4. *Geographic Mix*

During calendar years 1972, 1973 and 1974, the above-mentioned producer associations will endeavor to the fullest extent possible to avoid growth in shipments of steel mill products to any particular customs region of United States (Atlantic Coast, Gulf Coast/Mexican Border, Pacific Coast, Great Lakes/

Canadian Border, and off-shore regions) in excess of the approximate proportions of calendar years 1969, 1970 and 1971.

5. *Consultations*

The above-mentioned producer associations, through their authorized representatives, hold themselves ready to consult with representatives of the United States Government on any problem or question that may arise with respect to this voluntary restraint undertaking. They expect that, similarly, the United States Government would be prepared to consult with their representatives on any problem or question that may arise with respect to this voluntary restraint undertaking. They reserve the right to request consultation in the event that they consider they have been placed in a disadvantageous position with respect to other exporters of steel to the United States by developments in the international steel market taking place subsequent to entering this undertaking, or if developments in the international steel market should take place which could substantially impair the carrying out of this undertaking. Similarly, they recognize that the United States Government may request consultation if it considers that developments in the international steel market have taken place during the term of this undertaking which substantially affect any of the provisions of this arrangement.

The above-mentioned producer associations reserve the right to request consultation with respect to the exclusion in particular situations from the export limitation quantity in paragraph 1 above of shipments of large-diameter line pipe.

6. This voluntary restraint undertaking is based on the following assumptions:

a. The provisions of paragraphs 1, 2, 3, 4 and 5 of this letter should not have disadvantageous effects for the above-mentioned producer associations as compared with the undertakings made or to be made by producers of other countries concerning their exports to the United States market.

b. The United States Government will take no unilateral actions to restrict the quantity of steel mill products to be imported by the United States from Belgium, France, the German Federal Republic, Italy, Luxembourg, the Netherlands, and the United Kingdom, or to raise the tariffs thereon, including the new imposition of a surcharge, or to impose any other measure of similar effect.

c. This undertaking is made with the understanding that it is not in violation of any law of the United States or international rule, especially the provisions of the treaty establishing the European Coal and Steel Community or decisions of the European Economic Commission.

Sincerely yours,

This letter is signed on behalf of the under-mentioned associations by:

1. Comite de la Siderurgie Belge
   KSJ Baron Van Der Rest, President
2. Wirtschaftsvereinigung Eisen und Stahl Industrie
   Hans-Jorg Sendler
3. Chambre Syndicale de la Siderurgie Francaise,
   M. Jacques Ferry, President

4. British Steel Corporation
   Mark Littman
   Deputy Chairman on behalf of
   Lord Melchett, Chairman
5. British Independent Steel Producers Association
   Sir Douglas Bruce-Gardner, Chairman
6. Associazione Industrie Siderurgiche Italiane "Assider"
   Dott Mario D'onofrio, Delegato Della Presidenza
7. Groupement de l'Industrie Siderurgique Luxembourgeoise,
   M. Rene Schmit, President
8. Vereniging de Nederlandsche Ijzer en Staalproducerende Industrie,
   M. Evert Van Veelan, Chairman

May 4, 1972

Dear Mr. Secretary:

In order to assist in the maintenance of an orderly market for steel in the United States, the Japan Iron and Steel Exporters' Association (hereinafter referred to as the Association) declares its intention, with the cooperation of associations for steel mill products not covered by the Association, to extend through December 31, 1974, its limitations on steel mill product exports to the United States, as initiated by its memorandum of December 23, 1968. The following are the terms under which these exports will be regulated:

1. Total exports and growth rate. The total quantity of Japanese steel mill products to be shipped to the United States during calendar year 1972 will not exceed 5,895,000 metric tons.

The growth rate of exports of the aforementioned products will not exceed 2-1/2 percent per annum in each of the years 1973 and 1974 over the preceding year's limit of shipments. In accordance with this growth rate, such shipments will not exceed 6,042,000 metric tons during calendar year 1973 and 6,193,000 metric tons during calendar year 1974.

Shipments in either semester (6-month period) of a given year will be held to no more than sixty (60) percent of the total annual quantitative limit.

2. Product mix. Within the foregoing overall limitations, shipments of specialty steel mill products will be confined to total quantities not to exceed the following (in metric tons):

	1972	1973	1974
Stainless Steel	72,463	68,840	70,561
Tool Steel	893	915	938
Other Specialty Steel	155,935	141,901	145,448

With respect to the aforementioned specialty steel mill products, the Association will not depart significantly from the product mix in each calendar year as compared with the pattern of such shipments to the United States over recent years.

Since exports will be regulated by reference to Japanese customs clearance statistics, which are based upon the Brussels Nomenclature, Japanese classifications of the above grades of steel will be used, provided that in the case of tool steel a separate consideration will be given to the shipments in accordance with the definitions of the Tariff Schedules of the United States Annotated.

With respect to other categories of steel mill products, the Association will generally maintain the product mix in each calendar year in accordance with the average shipments for the period 1969, 1970, and 1971.

With special reference to exports of cold finished carbon steel bars, shipments will be confined during calendar year 1972 to a level of not more than 2-1/2 percent greater than during 1970. A growth rate of 2-1/2 percent of the preceding year's limit will not be exceeded during calendar years 1973 and 1974.

3. Fabricated structural steels. Such products are not considered steel mill products and are not within the limitations set forth in paragraph 1 above. However, with reference to shipments of fabricated structural steel products, as defined below, according to the definitions of the Japanese customs tariff—namely: electric poles (towers) and parts thereof, of iron or steel: steel towers and parts thereof, of iron or steel, n.e.s.: bridges and parts thereof, of iron or steel: flood gates and parts thereof, of iron or steel, n.e.s.: structures and parts thereof, of iron or steel, n.e.s.: bars, rods, plates, sheets, tubes, pipes and other materials, worked for structural use, of iron or steel—special efforts will be made to confine the shipments by Japanese steel producers during calendar year 1972 to a level of not more than 2-1/2 percent greater than during 1970, outside the quantitative limit as described in paragraph 1 above. A growth rate of 2-1/2 percent of the preceding year's limit will not be exceeded during calendar years 1973 and 1974. It is understood that problems relating to the shipments of fabricated structural steels may be reviewed at the end of 1972 in accordance with paragraph 5.

4. Geographic mix. During calendar years 1972, 1973, and 1974, the Association will endeavor to the fullest extent possible to avoid growth in shipments of steel mill products to any particular customs region of the United States (Atlantic Coast, Gulf Coast-Mexican Border, Pacific Coast, Great Lakes-Canadian Border, and offshore regions) in excess of the approximate proportions of calendar years 1969, 1970 and 1971. With specific reference to the Pacific Coast customs region, shipments will be confined to a level not exceeding one-third of total shipments of steel mill products to the United States.

5. Consultations. The Association through its authorized representatives, holds itself ready to consult with representatives of the United States Government on any problem or question that may arise with respect to this voluntary restraint undertaking. The Association expects that, similarly, the United States Government would be prepared to consult with the representatives of the Association on any problem or question that may arise with respect to this voluntary restraint undertaking.

The Association reserves the right to request consultation with respect to the exclusion from the export limitation quantity in paragraph 1 above of shipments of large-diameter line pipe (48 inch diameter or wider) when such pipe cannot be supplied from domestic production in the United States.

6. This voluntary restraint undertaking is based upon the following assumptions:

A. That, in consideration of the voluntary restraint undertaking in the existing memorandum of December 23, 1968, the terms of the undertaking of paragraphs 1, 2, 3, 4, and 5 above as well as the Association's share of total United States imports of steel mill products, will not be disadvantageous to the Association as compared with the undertakings made or to be made by the members of the European Coal and Steel Community.

B. That the United States Government will take no unilateral actions to restrict the quantity of United States imports of steel mill products, or to raise the tariffs thereon, including any new imposition of supplemental duties: and

C. That this undertaking is made with the understanding that it is not in violation of United States law or international law.

Yours sincerely,

/ s /

Yoshihiro Inayama
Chairman
Japan Iron and Steel
Exporters Association

During the first eight months of 1972, imports reached 10.8 million tons which is a substantial amount, and there is every indication that the annual figure will be between 16 and 17 million tons.

The nonquota countries, outside of Canada, have increased their tonnage considerably. For the first eight months it was 38% higher than the comparable period of 1971. Whether some action will be taken in the future to limit the tonnage from these countries will most probably be determined by the extent of this year's imports from them.

In his statement on the renewal of the voluntary restraint agreements, President Nixon expressed his approval.

Office of the White House Press Secretary
Saturday, May 6, 1972

The White House
Statement by the President

I am pleased to announce that the steel producers of Japan, the European Community and the United Kingdom have expressed their intention to restrain on a voluntary basis their exports of steel mill products to the United States during the next three years.

This welcome development—which will lead to the preservation of jobs in our domestic steel industry—successfully concludes more than a year of discussions.

For some time, I have been deeply concerned about the serious problem that excessive imports have posed for our steel workers and our steel industry. Consequently, I directed that efforts be made to limit the impact of steel imports.

The Deputy Under Secretary of State for Economic Affairs, Nathaniel Samuels, undertook discussions with the Japanese and European steel producers with respect to their renewal—on improved terms—of the voluntary restraints on steel exports.

After more than a year's effort, Mr. Samuels has succeeded in this important endeavor. The Secretary of State has advised me that he has received communications from the steel producers in these countries in which they pledge a three-year restraint—with improved terms—of their voluntary limits on their steel exports to the United States.

This undertaking represents a substantial improvement over the arrangements of the last three years and will enable domestic steel producers to make their plans with confidence that imports will not be disruptive in the domestic market, it will help preserve the jobs of American steel workers.

I am especially pleased that this undertaking was reached on a voluntary basis. Such statesmanlike cooperation is vital to our mutual efforts to build a more equitable and a more progressive system of international trading arrangements.

Opposition to the 1972 Agreements

Shortly after the voluntary restraint agreement had been extended, a suit was brought against it by the Consumers Union of the United States, Inc. This group, which described itself as representing approximately 350,000 members, sought to obtain an injunction to prevent the voluntary restraint agreement from operating. The defendants in this civil action were:

William P. Rogers
Secretary of State

Julius L. Katz
Deputy Assistant Secretary
of State for Economic
Affairs

United States Steel Corporation

American Iron and Steel Institute

The Japan Iron and Steel
Exporters Association

Nippon Steel Corporation

Fried. Krupp Huttenwerke A.G

British Steel Corporation

The Consumers Union maintained that the voluntary restraint agreement would cause the price of steel to go up, and that its members who purchased steel would be disadvantaged and the union itself, which is completing a building containing somewhat in excess of 170 tons of structural steel, would also be disadvantaged in this construction operation.

It should be noted that the defendants included, not only American companies, but foreign companies as well. It is also of interest to note that the importers of steel did not join in the suit. No disposition has been made of the civil action at this writing, and it remains to be seen what will happen.

4 The Steel Labor Agreement and Steel Productivity

One of the major influences affecting the steel business in 1971 was the bargaining on a new labor pact with the United Steelworkers of America (USW). As the industry's customers hedged against a strike, steelmakers had to contend with an extremely irregular pattern of domestic activity and sharply increased imports, an unwelcomed combination of circumstances that has become a tradition in times of labor negotiations now held every three years. The agreement signed on July 30, 1968 was scheduled to expire on July 31, 1971, and in view of interim developments on the labor front, both in steel and other industries, a strike appeared more likely than at any time since 1959. The wage and benefit increases won elsewhere by the USW and other unions made it a foregone conclusion that the price of avoiding a work stoppage would be an expensive steel settlement. On the other hand, steel companies were hard pressed financially, and domestic steel was in competitive difficulty vis-à-vis imports from lower wage countries. Such was the dilemma leading up to the 1971 labor contract deadline.

Events Shaping the 1971 Labor Negotiations

On October 1, 1970, with the two month strike at General Motors in progress, the USW held its annual convention in Atlantic City, New Jersey. Unemployment and inflation were underscored as major areas of concern, and with gains under the 1968 contract voided by rising prices, notice was served on the industry that a substantial wage increase would be sought in 1971 to redress this situation. Another announced objective for 1971 was a cost of living agreement to insure protection against inflation. Strike sentiment was strong and some local USW officials wanted the right to strike over local grievances. On November 11, 1970, the GM strike was settled with a first year wage increase of 51¢, a 3% wage increase based on productivity in the second and third years, and unlimited cost-of-living adjustments.

In mid-February 1971, the USW and National Can Corporation signed an agreement calling for a base rate wage increase of 50¢ per hour in the first year and $12.5¢ in each of the next two years, plus a cost-of-living escalator, improved pension benefits, and other improvements in fringes. The President's Council of Economic Advisers estimated that the resultant cost increases would amount to 13% in the first year and 9% per year over the life of the contract.

After a twenty-eight-day strike, the major can companies accepted the National Can Corporation settlement, which eventually set a pattern for the negotiation of USW contracts with the aluminum and copper companies. Once June and July saw these contracts settled for a 30% increase, or thereabouts, it became more and more likely that this would be the price of a new steel agreement.

On July 6, one day before steel bargaining was to begin, the negotiators, upon request from the White House, met with President Nixon who stressed the need for a constructive settlement, namely, one that would help make the industry more productive. A Cabinet steel study prepared by Hendrik Houthakker of the Council of Economic Advisers had detailed the extent of the profit squeeze in steel and the substantial inroads made by foreign competition. Clearly, these problems would be aggravated by a work stoppage, but then again, a substantial pay boost would also worsen the industry's competitive position. Furthermore, both a strike and the upward influence on prices from an expensive settlement would work contrary to the administration's efforts for a general economic recovery. At the July 6 meeting, the president did not suggest any percentage or dollars and cents terms, but administration officials, aware that steel bargaining would likely follow the pattern set in the can agreement, suggested that an expensive contract would only be considered "constructive" if backed up by measures to improve steel productivity.

The Nine Company Contract Talks

On July 7 negotiations were started at the Shoreham Hotel in Washington, D.C. Significantly, the industry group had been reduced to nine companies by the withdrawal of Wheeling-Pittsburgh and CF&I Steel. Both these companies, as well as a number of others, sought to protect themselves against a strike by obtaining contract extensions which included the pledge to sign the same agreement which was reached by the negotiating companies. Thus they would have the advantage of operating while the others were on strike, and this was considered necessary because of the financial state of the companies. Extensions were actually obtained by Alan Wood Steel Company, Lukens Steel Company, and the Portsmouth plant of Cyclops. Wheeling-Pittsburgh attempted without success to negotiate a contract extension to accept in advance the terms generally agreed to by the industry.

In addition to the extensions, there were a number of companies whose contract had been negotiated to expire two months after the normal contract, and this gave them assurance of operating for the next 60 days if a strike took place on August 1, 1971. These included Kaiser Steel Corporation, and Granite City Steel Company. Further there were some plants of the major companies, such as National Steel Corporation and Armco Steel Corporation, which were organized by a union other than the USW and, therefore, were not subjected to the possibility of a strike on August 1, 1971.

A number of other steel producers, particularly the mini-mills, were considered immune from a strike on August 1, 1971. Consequently, even if a strike was called, steelmaking capacity, which was estimated at some 37 million tons, or 24% of the nation's total, would continue to operate.

Further, there was a large supply of imported steel available. Add this to the fact that a number of consumers had stocked inventory during the first six months, and it was easily seen that a steel strike would not cause a shortage for quite some time. It should be noted, however, that the percentage of the steel industry operating would not have been able to provide the complete product spread necessary for the economy to function.

An industry-wide strike had been ruled out, and it was evident that the no strike concessions made by the union to some companies reflected concern—particularly within the top union leadership—about the industry's import, profit, and productivity problems. The conflict between the union and the industry was set forth well in a comment made at the meetings held in Washington by both sides in mid May. After the industry had detailed its financial position, which was not particularly strong, one of the union members remarked to an industry representative "You ain't got it but we've got to get it." With this atmosphere it seemed that the union and the industry were on a collision course, unless the demanded substantial wage increase was granted and accompanied by a considerable price increase.

The nine company bargaining team was headed by R. Heath Larry, vice chairman, U.S. Steel and included John J. O'Connell, vice president, Bethlehem Steel; J. Warren Shaver, vice president, labor relations, U.S. Steel; and John R. Wall, vice president, personnel, Republic Steel. In addition to I.W. Abel, president of USW, the top union group included Secretary-Treasurer Walter J. Burke and Vice President Joseph P. Mahony.

For the most part, the first three weeks of bargaining were taken up with the resolution of local issues, and with the July 31 deadline only days away, the major money problems were still unresolved. On July 29 the company group made a firm offer, fundamentally patterned after the can settlement, and during the next two days, the two sides concentrated on ironing out the complex details and mechanics of a new contract. By the morning of July 31, it appeared that an accord was imminent. However, as the midnight strike deadline approached, some disagreement over incentives and a number of problems in contract language remained, and with just hours remaining on the 1968 contract, USW president Abel received approval for a twenty-four-hour extension from the union's 500 member steel conference. By 2:00 A.M. on August 1, 1971, a tentative agreement had been reached. It received unanimous union ratification later in the day, and when signed by the USW and the nine-company group, 350,000 steelworkers had a new contract, which set the pattern for subsequent bargaining throughout the industry.

Terms of the 1971 Agreement

As a result of the three-year pact, which was to run to August 1, 1974, the nation's steelworkers received substantial increases in wages and fringe benefits. On major contract items, the agreement followed the pattern previously set in the automobile, can, aluminum, and copper industries. Since it became effective August 1, its first year increases in wages and benefits were not affected by the ninety-day wage-price freeze instituted on August 15. The major provisions of the contract are as follows.

Cost-of-Living Adjustment. Commencing August 1, 1972 and at three-month intervals thereafter, a cost-of-living adjustment shall be added on to the employee's pay at the rate of 1.0¢ an hour for each 0.4 of a point increase in the Consumer Price Index published by the U.S. Department of Labor, Bureau of Labor Statistics. The index for June 1972 represents the base used to calculate these adjustments. If the Consumer Price Index falls below the base index, there shall be no cost-of-living adjustment, in which case provision is made for "guarantees" amounting to at least 25¢ an hour over and above the other stipulated wage increases by the expiration of the contract.

Productivity Committees. At the outset of the agreement under section 1, "Purpose and Intent of the Parties," provision is made for a joint labor-management advisory committee on productivity at each plant. This new dimension in steel labor relations is detailed in paragraphs 1.3 through 1.6 of the contract, which read as follows:

> The parties recognize that for their joint benefit, increases in wages and benefits should be consistent with the long-term prosperity and efficiency of the steel industry.
>
> The parties are concerned that the future for the industry in terms of employment and return on substantial capital expenditures will rest heavily upon the ability of the parties to work cooperatively to achieve significantly higher productivity trends than have occurred in the recent past. The parties are acutely aware of the impact upon the industry and its employees of the sizable penetration of the domestic steel market by foreign producers. The parties have joined their efforts in seeking relief from the problem of massive importation of foreign steel manufactured in low-wage countries. Thus, it is incumbent upon the parties to work cooperatively to meet the challenge posed by principal foreign competitors in recent years. It is also important that the parties cooperate in promoting the use of American made steel.
>
> In order to implement this expression of purpose, a joint advisory Committee on Productivity shall be established at each plant. The Union representation on this Committee shall be two members of the Union in addition to the Local Union President and the Chairman of the Grievance Committee. The Union members shall be certified to plant Management by the Union and a corresponding number of Company members shall be certified to the Union. The Company and the Union members of the Committee shall meet at mutually

agreeable times, but no less than once each month. The function of the Committee shall be to advise with plant Management concerning ways and means of improving productivity and developing recommendations for stimulating its growth so as to promote the purpose of the parties as set forth above and also promote orderly and peaceful relations with the employees, to achieve uninterrupted operations in the plants, to promote the use of domestic steel and to achieve the desired prosperity and progress of the Company and its employees.

The General Superintendent or the Local Union President may from time to time suggest to the Committee areas of special concern consistent with the purposes of this Committee and the provisions of this Agreement. The functioning of this Committee shall not affect the existing rights of either party under any other provision of this Agreement.[1]

Rates of Pay. The contract increases steelworker wages 31% over three years. It provides for an across the board increase of 50¢ an hour during the first year and 12.5¢ an hour during each of the next two years. It increases the increments between each of the thirty-three job class wage rates by 0.5¢ an hour during the first year and 0.4¢ an hour during the second and third years.

Added Holiday. Beginning with the year 1972, the day after Thanksgiving Day is specified as a holiday.

Grievance Procedures. The contract streamlines procedures for the adjustment of complaints and grievances.

Other Contract Provisions. Contract provisions governing employee safety and health are expanded; the contract provides that beginning in 1972, vacation bonuses are to be established on a sliding scale to encourage vacations at times other than the summer months and the Christmas season; pensions are improved effective August 1, 1972; supplementary unemployment benefits are increased; the Sunday premium rate is increased from time and one-quarter to time and one-half effective August 1, 1973; provision is made for increased major medical and life insurance coverage.

The Emphasis on Steel Productivity

Insofar as the economic relationship between wages and productivity is concerned, the 1971 steel labor agreement was an unusual document. Wage increases that exceed productivity gains are inflationary, so that a noninflationary steel settlement, given the flat trend in steel productivity during 1965-70, would have entailed virtually no wage hike at all. Consequently, productivity was not used as a basis for wage negotiations, and attention was focused on the increased cost of living and the agreements previously negotiated by the USW in the can, aluminum, and copper industries. Having settled on a 31% wage increase over three years and on a previously eliminated cost of living escalator, the USW and the steel companies then agreed to establish plant level committees to work at raising the rate of productivity growth.

The contractual emphasis on steel productivity came after a period of record spending for plant and equipment designed with increased productivity in mind. During 1965-70, steel producers invested over $10 billion in new facilities, but steel productivity gauged by the output per man-hour measurement of the Bureau of Labor Statistics, showed little growth (0.4% annually). This certainly requires an examination of the latest developments in steel technology, the substantial amounts of capital committed to their installation, and the very concept of productivity itself. The question that has been raised is:

> With an all-time high in spending for new BOFs, continuous casters, and computerized rolling mills, all designed for faster, more efficient steel production, why has steel productivity failed to increase substantially?

To develop an adequate answer to the question, it is of primary importance to realize that the widely used indexes of output per man-hour provide, at best, a partial productivity measurement, which is crude in many respects and prone to misinterpretation, particularly when applied to individual industries, companies, and plants.

A relatively flat trend in an industry's index does not necessarily mean that its workers are not working as they should, that management is not managing, or that adequate, well-planned investments in efficient plant and equipment have not been made. It is also important to realize that international productivity comparisons, in which the United States has fared poorly of late, frequently are not qualified as to their shortcomings and, consequently, tend to exaggerate output per man-hour improvements in other nations.

This is not to deny that a productivity problem exists, either nationally or in the steel busines, or that domestic producers have all but lost a once comfortable advantage in output per man-hour levels. However, the significance of productivity to the nation's economy and its important role in Phase 2 of the Nixon Administration's wage-price policy raises some very serious questions about the concept, its measurements, and the reasons why its improvement has become difficult. The concept is surrounded by considerable confusion, even resentment, and in view of the renewed attention it is now receiving, a discussion of productivity problems, particularly as they relate to the steel industry is very much in order.

Output per Man-Hour in Steel

Some idea of the productivity problem, both in the steel business and in the overall economy, is provided by the indexes of output per man-hour published by the Bureau of Labor Statistics (BLS) for major sectors of the economy and for selected industries. They indicate steel's lag in productivity growth has been more pronounced than that incurred by the total private economy and by manufacturing industries in general.

To begin with, the BLS reports that the average yearly growth rate of output per man-hour in the steel industry since 1947 has been 1.7%, compared with a 2.9% rate for all manufacturing industries and a rate of 3.2% for the total private economy. During the first half of the 1960s, the annual growth rate of output per man-hour in steel pretty much kept pace with that for all manufacturing. It improved markedly to 4.3% well above its annual rate of less than 1% over the 1950s.

However, during 1965-70, while the growth rates for manufacturing and the total private economy both slipped to 2.0%, steel's rate incurred a much sharper decline, averaging a minimal 0.4% annually over the period.

According to BLS data, long-term rates of productivity growth in the steel industry have been lower than in most other industries that constitute the total manufacturing sector. In its productivity measurement program, the Bureau computes output per man-hour indexes for twenty-seven manufacturing industries, which are ranked in Table 4-1 according to their average annual productivity growth rates during 1957-70.

The 3.2% advance in total manufacturing productivity over this thirteen-year period reflected diverse rates of industry improvement, which ranged from a low of 0.4% in the footwear industry to a high of 6.3% in petroleum refining. Steel's growth rate of 2.2% placed it twenty-fifth out of the twenty-seven industries covered.

While positive growth rates have prevailed throughout the manufacturing sector over the long-term, twelve of the industries listed below experienced actual reductions in output per man-hour during 1970. These included the two largest manufacturing industries measured, motor vehicles and steel, which posted productivity declines of 6% and 3% respectively.

In contrast to the United States experience during 1965-70, BLS studies indicate that other nations have registered sizable gains in their productivity growth rates and have substantially reduced the output per man-hour advantage long held by American producers. International comparisons published by the BLS for the manufacturing sector are provided in Table 4-2, and the growth of output per man-hour in the United States is shown to be the lowest among the major industrialized nations of the free world.

Of particular significance is the wide gap between the United States and Japanese growth rates. The table also shows that the greater increases in productivity abroad substantially offset greater advances in hourly compensation rates, and consequently, unit labor costs rose faster here than in every other country except Canada.

In steel, according to BLS data, our productivity advantage which was substantial at the start of the last decade, has almost disappeared. In 1960 steel

Table 4-1
Growth of Output per Man-House in Selected Manufacturing Industries (Average Annual Rates of Growth)

Industry	1957-70
1 Petroleum Refining	6.3%
2 Hosiery	6.3
3 Malt Liquors	6.2
4 Radio and Television Receiving Sets[a]	6.0
5 Major Household Appliances[a]	5.8
6 Cigars	5.5
7 Aluminum Rolling and Drawing[a]	5.2
8 Flour and Other Grain Mill Products	5.1
9 Man-Made Fibers	4.7
10 Cement, Hydraulic	4.6
11 Tires and Inner Tubes	4.6
12 Sugar	4.5
13 Bottled and Canned Soft Drinks[a]	4.3
14 Paper, Paperboard, and Pulp Mills	4.2
15 Primary Aluminum	3.8
16 Concrete Products[a]	3.7
17 Motor Vehicles and Equipment	3.6
18 Corrugated and Solid Fiber Boxes[a]	3.2
19 Canning and Preserving[a]	3.0
20 Tobacco Products–Total	3.0
21 Candy and Other Confectionary Products	3.0
22 Glass Containers	2.5
23 Gray Iron Foundries[a]	2.4
24 Primary Copper, Lead, and Zinc[a]	2.2
25 Steel	2.2
26 Cigarettes, Chewing and Smoking Tobacco	1.6
27 Footwear	0.4

[a]Rates of increase for period slightly different than indicated.
Source: U.S. Department of Labor, Bureau of Labor Statistics.

producers in the United States utilized 15.0 man-hours per ton of steel mill products shipped, much less than 24.2 man-hours required in West Germany and the 44.2 man-hours for the Japanese industry. By 1965 man-hour requirements per ton in this country had declined to 12.2, still significantly lower than in West Germany (21.6) and Japan (27.9). However, by 1970, man-hour requirements in the three countries were moving close to parity; in the United States, they had declined only slightly to 11.9, while in West Germany and Japan, they had been reduced to 14.5 and 12.4, respectively.

Table 4-2
Growth of Output per Man-Hour, Hourly Compensation and Unit Labor Costs for all Employees in Manufacturing, Eleven Countries, 1965-70 (Average Annual Rates of Growth)

	Output per Man-Hour	Compensation per Man-Hour	Unit Labor Costs[b]
Japan	14.2%	15.1%	0.8%
Netherlands	8.5	11.1	2.5
Sweden	7.9	10.6	2.5
Belgium	6.8	8.4	1.4
France	6.6	9.5	2.7
Switzerland[a]	6.2	6.2	0.0
West Germany	5.3	8.7	3.2
Italy	5.1	9.1	3.8
United Kingdom	3.6	7.6	3.8
Canada	3.5	8.3	4.6
United States	2.0	6.0	3.9

[a]Wage earners only.
[b]Rates are computed on the basis of national currencies.
Source: U.S. Department of Labor, Bureau of Labor Statistics.

The Concept of Productivity and Its Measurement

It must be reiterated that the findings just reviewed are based on a partial analysis of productivity, whereby the BLS gauges output in relation to only one of the input factors of production, namely, manpower, expressed in terms of man-hours. The reasoning behind the Bureau's approach is discussed below, and to appreciate its limitations and their effect on the results obtained, the concept of productivity and some of the more important aspects of its measurement must be clearly understood.

Productivity Is Input-Output Relationship. Productivity is expressed as a ratio or relationship between the *input* factors in production and the resultant *output* or product. It is fundamentally the production obtained per unit of resources consumed in the productive process, and its improvement generally means that more has been obtained from a given input. Consequently, the concept of productivity is interwoven with the most basic of economic objectives: the optimum utilization of scarce resources.

To increase productivity means to take what we have and make more of it, more goods and services to raise living standards, and the rate of productivity growth, therefore, indicates how well producers have used the various inputs or productive resources to satisfy the needs and wants of consumers. In short, an increase in productivity is an increase in real wealth.

Input Consist of Three Categories. The input factors in any manufacturing operation can be grouped into three general categories: manpower, materials, and machinery. These inputs or resources function together, with manpower applying the machinery to materials in order to shape some product. This makes it difficult, if not impossible to determine with precision what portion of the final output is attributable to each of the three interrelated input factors.

An increase in productivity generally reflects the joint influence of many sources of improvement in resource utilization, such as an increase in the productiveness of labor, upgraded management skills, the successful introduction of new technology, an increase in capital investment per worker, a closer to optimum rate of operations, a more efficient plant layout and flow of production, and increased use of upgraded raw materials.

Manpower Input Difficult to Measure. The general input category, manpower, encompasses the total human effort, mental and physical, required to sustain a particular manufacturing operation. In most industries, this "all employee" input is difficult to measure, since the types of employment utilized are numerous and are characterized by a wide variety of skills. The steel industry, for example, employs production workers in thirty-three different job classifications, as well as nonproduction workers in diverse managerial, administrative, supervisory, professional, technical, and clerical positions.

Manpower requirements are most often quantified as "man-hours" worked, a more suitable measure than employment statistics, which often do not distinguish between full-time and part-time employees and do not account for changes in the duration of the work week as business activity expands or contracts. Moreover, it is important to recognize that man-hours are not homogeneous. To fail to do so is to make the invalid assumption that all man-hours contribute equally to the productive process, or in other words, to contend that the hours of skilled workers and management personnel are no more significant than the hours of unskilled workers.

One hour worked by the roller on a hot strip mill, for example, certainly makes a more substantial contribution to steel production than one hour worked by a clean-up man. This is not to denigrate the work of the latter, which is necessary or it would not be performed. Nevertheless, recognition of the superior contribution of the roller is made in his rate of pay, which is set higher than that of the unskilled worker.

Steel Material Inputs Vary Widely. The second input category, materials, is comprised of all the goods and services needed to turn out the final product. In the steel industry, raw materials, such as iron ore, coal and limestone are used, as well as processed materials, including beneficiated iron-bearing materials and coal, scrap, alloys, refractory brick for furnace and ladle linings, electricity to operate the mills and furnaces, and a number of other material inputs.

The raw materials for steel production are found in nature, and depending upon the quality of the deposits used, steel productivity can be improved or adversely affected. In a particular blast furnace plant, for example, much better results can be achieved by charging iron ore from Australia (64% Fe), as opposed to that from either the Mesabi Range (49% Fe) or from the Birmingham district (36% Fe).

Over the last decade, however, as steel producers have increased their use of beneficiated or upgraded iron-bearing materials, the influence of direct shipping ore on blast furnace performance has become less significant than it once was.

Beneficiated Ores Becoming Standard. The consumption of beneficiated ores (pellets, sinter, briquettes, and nodules) in the nation's blast furnaces increased from 57.8 million net tons in 1960 to 109.0 million net tons in 1970, the latter figure representing three-fourths of all the iron-bearing materials consumed, up from somewhat more than one-half ten years earlier. In fact, many blast furnaces now use a 100% treated burden.

This is significant from the standpoint of productivity measurement, because once beneficiation has been accomplished, the blast furnace no longer uses a raw material, but one that has had capital and labor applied to it. The same holds true in the coke-making process, wherein beneficiated coal has been charged in ever increasing quantities.

As a result, the material inputs for steel production have become more nearly comparable to those used in other manufacturing industries, since they are, in fact, a manufactured product.

Advances in Steel Technology Very Important. The third input category, *machinery*, is of primary importance, for it is the principal means by which production has been increased in quantity and improved in quality during the past half century or more. The advances in steel technology, particularly since the close of World War I, have made it possible for steel producers to supply a significantly greater market with much less manpower. In 1919, for example, when the average workweek in the steel industry approached sixty hours (ten to twelve hours per day over a six-day week), U.S. Steel shipped 13.5 million net tons of steel products and employed 252,106 persons. Fifty years later, in 1969, its shipments were up 66.3% to 22.4 million net tons, even though employment had declined 18.8% to 204,723 persons and the workweek had been shortened to less than forty hours (seven to eight hours per day over a five-day week).

A number of difficulties are involved in quantifying the machinery input in terms of the fixed capital investment needed to sustain production. Because of fluctuations in the value of money, complications arise in making an accurate evaluation of the facilities in question and also in determining the extent to which they have already been written off. This is particularly the case in industries such as steel, which operate heavy, long-lived equipment, giving price changes of significant magnitude time to occur.

Although productivity measurement should be based on the economic value of the plant and equipment employed, most company accounts value assets only on an original cost basis. If such values are used to quantify the machinery input the consumption of capital per unit of output is understated, and consequently, productivity is overstated.

True Steel Industry Output Difficult to Measure. If a firm or industry produced a single, uniform product, there would be little difficulty in comparing its current output with that in a prior period. Complications arise in the case of multiple product lines to which new products are added and certain others are dropped over a period of time, and also when significant quality improvements have been introduced.

The steel industry's output is not homogeneous, but consists of a wide variety of products of various shapes, sizes, chemistries, and finishes. The frequency of alterations to this complex product line is indicated by the fact that the United States Steel Corporation makes one new product introduction per week.

Simple Tonnage Figures Not Satisfactory. In expressing the industry's output in physical terms, some means must be employed to account for changes in its composition, particularly in view of the extensive shift during recent years to the production of high-strength, light-weight steels, which reduce demand requirements and output on a tonnage basis. Obviously, if steel output is quantified by means of tonnage statistics, which are not weighted in some manner to reflect such changes, a distorted measurement of steel productivity is obtained.

There are several possibilities for measuring a diversified output, such as expressing it in terms of money value or value added. A common characteristic of these measures is that each entails a degree of aggregation, for the diverse product line must in some manner be unified on some common basis. The underlying assumption in using dollar values to aggregate varied physical outputs is that the more sophisticated, high quality products carry higher price tags, or that their dollar values reflect the additional resources required in their production.

Even Dollar Values May Be Inaccurate. In the steel industry, however, the same competitive pressures that have induced the production of lighter, higher quality steels in recent years have also prevented their additional production costs from being fully reflected in their prices or dollar values. Consequently, if steel output is aggregated in terms of dollar values, it is understated relative to the consumption of resource inputs, and steel productivity is there by given a downward bias.

Output Per Man-Hour as a Productivity Measure

Although there are three input factors in any manufacturing operation, namely, manpower, materials, and machines, the accepted and most widely applied productivity measure relates output to only one of the input factors—manpower. As noted previously, increasing or decreasing productivity is determined by the BLS in terms of output per man-hour, or conversely, man-hours per unit of output.

This partial productivity ratio involves a number of significant shortcomings. Since it excludes the two input factors, materials and machines, their productive efficiency can only be evaluated indirectly in terms of their effect on man-hour output. In certain instances, this can be grossly misleading; it is quite possible that the installation of a new piece of equipment may double man-hour output, yet the equipment may cost much more than twice its predecessor in constant dollars. As a result, labor productivity in output per man-hour terms would increase, but capital productivity would actually decline. In like manner, increases in labor productivity obtained by using higher quality production materials frequently have been offset by the capital costs involved in upgrading these materials.

BLS Figures Overlook Quality Improvements. While BLS output per man-hour indexes tend to overstate long-term productivity trends by considering only one of the input factors, they tend to understate productivity by measuring output in quantitative terms, thereby overlooking quality improvements.

The downward bias to productivity becomes more significant as efforts to improve product quality are intensified. Such efforts increase the input of capital and man-hours, without a corresponding increase in quantitative output. In this regard, it must be noted that a considerable improvement in the quality of steel products has been made over the last decade.

Confronted with these difficulties, one may ask why the BLS continues to use partial productivity ratios of output per man-hour as the principal means of productivity measurement. One reason is that a primary concern of the Bureau is productivity in the economy as a whole, and here the labor factor is much more significant than it is in a particular industry.

Individual industries invest in plant and equipment, the cost of which stems in great part, from a considerable labor input used by equipment builders, and they, in turn, buy other capital goods containing more labor, etc. Therefore, when the entire economy is under consideration, man-hour output is a highly significant figure, with labor input accounting for as much as four-fifths of total production costs.

On the other hand, when one is concerned with the productivity of an individual industry, and this is required under the Phase 2 stabilization program, the situation is different, because the purchase of plant and equipment by an individual industry represents capital investment. Naturally, this shortcoming is most serious in capital intensive industries such as steel.

Output per man-hour indexes for the steel industry, by neglecting to relate steel output to the substantial machinery input used in its production, tend to present an exaggerated trend in productivity growth, especially since the capital investment required to increase man-hour output often has been much greater proportionately than the resultant increase in labor productivity. The following example will illustrate this point.

Consider a New Rolling Mill. A primary rolling mill, which takes a basic ingot and rolls it down into a semi-finished product, such as a slab or bloom, was replaced in 1961 at a cost of approximately $51,000,000. The crew on the new mill was virtually the same as that on the previous mill, which rolled steel at a rate of 1500 tons per turn. The output of the new facility was virtually double that of the old one, so that on the surface, if man-hour output is used as the measurement criteria, it would appear that productivity had been doubled.

However, when one considers that the investment of $51,000,000 in the new mill was made to replace a mill installed in 1912 at a cost of $2,000,000, it is obvious that the amount of capital involved, even if one makes corrections for the purchasing power of the dollar, was much greater than the amount invested in 1912.

An estimate based on construction cost indexes indicates that the 1961 mill, in terms of 1912 prices, could have been installed for approximately $15,000,000. This was over seven times the amount invested in 1912, and although the 1961 mill was far superior to its 1912 counterpart, the additional capital cost involved far outweighed the resultant twofold increase in output.

The use of man-hour statistics for individual industries has often been defended on the ground that such statistics are available and are much more homogeneous and more easily handled than data on the other input factors. In response to this, one might note that the only homogeneous factor in man-hours is the hour, since the quality of labor, both physical and mental, varies over a wide range. Nevertheless, in computing its output per man-hour indexes, the BLS treats all man-hours the same, or assumes that all types of manpower contribute equally to the final output.

BLS Qualifications Often Ignored. In publishing output per man-hour data for the steel industry, the BLS has gone to considerable lengths in explaining the concept used, its limitations, and the technological developments in the industry that have helped shape the output per man-hour trends reported. The diversity of steel output is provided for by an elaborate system of weights, whereby the various steel products are differentiated into more than 130 categories. However, this complex method of weighting output statistics does not account for the recent shift to lighter weight steels. The bureau has stressed that its indexes do not measure the specific contribution to steel production made by labor, capital, or any other productive factor, and has cautioned that they should be taken only as general indicators of productivity trends in the industry.

In practical use, however, the bureau's emphatic qualifications have often been ignored, and its indexes have been treated as exact indicators of steel productivity.

Shortcoming of International Productivity Comparisons

Direct comparisons of output per man-hour trends among countries are very often deceiving, if not invalid, and this applies particularly to the steel industry. For example, in Japan a number of the man-hours used in the steelmaking process are not counted in gauging productivity. The Japanese have a practice of contracting-out a large share of their maintenance, as well as their raw materials handling, and the actual shipment of finished steel products and the man-hours used for these purposes are not reflected in published employment statistics.

In one of their major plants, almost half of the worker-force of 7,000 people is supplied by outside contractors. In a number of instances, 80% of the maintenance is contracted-out, and man-hours used at the coke ovens in some plants are assigned to nonsteel divisions within their parent companies, rather than to steelmaking operations.

U.S.-Japanese Productivity Comparisons Not Valid. Given such practices, the Japanese rate of steel productivity, although it has been increasing, cannot be compared to that in the United States, where the man-hours employed in maintenance, raw materials handling, and shipments are part of labor input.

A special arrangement between a Japanese steel company and a German firm sought to develop reasonable productivity comparisons for somewhat similar steel plants in the two countries, but even here, despite the attempt to establish comparable standards, problems arose to make exact comparisons difficult and, in some cases, impossible.

U.S.-European Comparisons Also Difficult. When steel productivity in the United States is compared with that in European countries, the problem of determining the exact number of man-hours utilized is again serious. The German steel industry has even had difficulty in making comparisons among companies in Germany itself because of the inclusion or exclusion of certain operations in the overall steelmaking process.

Further, it is an established fact that European mills are production oriented to a greater extent than American mills; they accumulate orders over a much longer period of time to permit the most economical rolling schedules. This allows for a higher degree of productivity at the expense of customer service.

American mills, on the other hand, because of a higher degree of market competition, are customer oriented and provide more rapid delivery by not accumulating orders over as long a period as their European counterparts.

Although every effort is made to establish efficient rolling schedules, service to the customer nearly always takes precedence. This factor, among others, reduces the validity of international productivity comparisons, which are made on a straight, unqualified, man-hour output basis.

Reasons for the Lag in Steel Productivity

The steel industry's experience since 1965 indicates that spending for new facilities, however massive, does not result in any automatic or immediate increase in productivity as it is measured. Rather, a number of difficulties must be solved and certain preconditions must exist before the full benefits of new plant and equipment can be realized. The reasons why productivity improvement in the steel industry has been difficult to achieve include the following.

Inadequate Volume and Steel Imports. With the domestic steel market's growth absorbed in great part by foreign producers, domestic shipments have remained relatively static since 1965, and because of inadequate volume, the industry's new facilities have not been operated at their potential on a sustained basis. By contrast, the recent experience overseas, particularly in Japan, has demonstrated the relationship between a growing volume of production and increased steel productivity. If sufficient tonnage is put through the modern equipment now operating in this country, new productivity records can be set.

Scheduling Problems and Steel Imports. In addition to the volume of steel production, the manner in which this production is scheduled bears an important relationship to productivity. If the schedule for a hot strip mill, for example, can be set to permit the rolling of standard, large-size slabs into large coils over an extended period of time, let us say four hours, productivity will be high. If, however, orders cannot be grouped for efficient production runs during the same four-hour period, and slabs in a variety of sizes and chemistries, including small slabs for small coils, must be scheduled for rolling, productivity will be substantially lower.

In recent years, steel productivity has suffered because steel imports have been concentrated in stocked sizes where demand is greatest. This has removed a substantial amount of high-productivity tonnage from domestic mill schedules, leaving a greater proportion of difficult, less efficient rollings.

Lighter Steels Reduce Tons Per Man-Hour. As competitive pressure and market demand have required the production of lighter steel products, productivity, measured on the basis of man-hours per ton, has been adversely affected. Tons are a varying norm for measuring output, insofar as lighter steels can satisfy a specified market requirement or perform a given function with less tonnage than their heavier counterparts.

A good example of this is provided by the shift to double-reduced or "thin" tinplate. It is customary for tinplate to be sold by the base box, which represents an area of so many square inches. In 1960 the base box weighed about 92 pounds, but currently, because of the introduction of double-reduced tinplate, a high percentage of base boxes weigh 55 pounds.

This means that the same number of cans is now produced from 40% less steel, measured on the basis of weight. Consequently, although the tonnage of tinplate production has grown during the last five years, output per man-hour in this segment of the industry, measured in terms of tons, would have improved to a much greater extent, if the base box had remained at 92 pounds.

The same phenomenon has taken place in the production of structural steels. In this instance, sections are now lighter than formerly, and it has been estimated that the amount of steel used in the Empire State Building, some 53,000 tons, could be used today to construct a building 13 stories higher. Examples such as this could be multiplied; the point is that the steel industry is now producing lighter steels, so that indexes of man-hours per ton, with emphasis here on the ton, have become somewhat obsolete as gauges of steel productivity.

Stress on Quality Output. The substantial upgrading of steel quality in recent years has been due, in great part, to improvements in the process of production and the equipment involved. However, additional hours for inspection have been added in many instances, and these man-hours do not produce an extra pound of steel. In fact, when inspection is severe, some tonnage of steel is rejected. With more man-hours applied by virtue of rigid inspection, fewer tons of steel frequently are shipped.

High Technological State of Industry. Another reason why productivity increases in steel have become difficult to obtain is the high degree of technology already found throughout the industry. The 1965 figure of 12.2 man-hours per ton of shipments was good to start with, and it has not been an easy matter to improve it. The same situation is rapidly developing in Japan and other industrialized countries where marked increases in productivity have been recorded over the past decade. At the present time the steel industries in these nations are moving up against the limits of technological development, and without question a slowdown in their rate of productivity increase can be looked for in the near future.

Extensive Construction Time. Given their size and complexity, construction projects for new steel facilities have required years to complete, and consequently, it has been customary for considerable time to elapse between capital expenditures and the start of regular operations, when the possibility of any productivity benefit arises for the first time.

Prolonged Break-in Periods. The highly sophisticated, computerized steel facilities installed in recent years have required longer than anticipated break-in periods, extending the delay of regular operations and possible productivity improvement.

Difficulty of Maintenance and Repair. Despite their tremendous potential for efficient, high-quality production, the extent to which new steel facilities rely on highly complex and technical electronic devices has increased maintenance requirements and the possibility of breakdowns and has made repairs more difficult than on older, less intricate equipment. Consequently, outage time on new facilities is greater, a disadvantage where productivity is concerned.

Continuous Casting Problems. Among the most recent developments in steel producing technology, continuous casting stands out as a genuine breakthrough, affording a great potential for marked productivity increases. Unfortunately, however, implementation of the process on a large scale has encountered considerable difficulty, and the sizable investments it has attracted over the last decade have produced mixed results. Although some types of billet casting have become a routine steelmaking operation, the products rolled from this semi-finished shape constitute a very small percentage of the industry's output.

On the other hand, a frustrating variety of technical and operating problems has slowed the casting of wide slabs, which are required to roll high tonnage steel products, such as auto body sheets and tinplate. Following extensive delays in development, construction, and start-up, a number of commercial-wide slab casters have attained an acceptable degree of success, while other installations continue to have their problems, even though years have passed since they were first put into operation. Consequently, the productivity potential of continuous casting investments has not as yet been fully realized.

Need to Balance Facilities. Productivity improvement is curtailed when both new and relatively obsolete facilities are in operation in the same plant. This tends to produce a series of bottlenecks, so that output from the new facilities is limited by the speed and efficiency of the older ones.

In the case of one steel company, for example, a new basic oxygen shop has been erected at a cost of $75 million, and a new hot strip mill has been installed at a cost of $125 million. In between these two new facilities, a slabbing mill, which has been in operation for some time, is used to reduce the plant's production of basic oxygen steel ingots to slabs, which are then processed into hot-rolled coils on the strip mill.

Although the slabber is in good operating condition, its production capabilities are out of balance with those of the new BOF shop and strip mill, and consequently, it is limiting the output of the entire plant. A new, more powerful slabbing mill or a continuous caster would result in more production and higher productivity.

Similar examples, which indicate the problem of integrating new technologically superior facilities into the production schemes of long-established steel plants, can be cited for other steel companies in the United States and Europe. It should be noted, however, that the Japanese steel companies have, for the most part, been developing totally new plant sites, and therefore productivity benefit has been more readily obtained from their capital investments.

Pollution Control Expenditures. Steel productivity has been adversely affected by industry-wide pollution control programs, which have increased capital costs and manpower requirements without any corresponding increase in steel output. In recent years, spending by steel companies for air and water pollution controls has accounted for a growing proportion of their total capital expenditures, and to date, such spending has aggregated over $1.3 billion, more than in any other major industry.

Rising Capital Costs and Interest Rates. In recent years, sharply inflated construction costs and record interest rates have reduced potential for increased productivity per dollar of capital investment in new steel facilities.

Incentives on New Equipment. When new steel facilities are installed and labor incentive rates must be established, a traditional contest arises between labor and management. Labor would like to see the incentives established after a relatively low tonnage rate is achieved, while management would prefer to have them set after a higher tonnage output is achieved. This is often the reason why new equipment, which represents a substantial investment, does not begin to yield the returns it should in terms of productivity until after incentive rates have been established, and in some instances this could take almost a year.

Labor Market Conditions. In recent years, steel productivity has suffered because of labor market conditions in a number of the major steel producing areas, most notably Chicago, Detroit, and Cleveland, where a rapid increase in turnover rates has required expanded training programs for new employees.

Prospects for Increased Steel Productivity

The prospects for increased steel productivity in the years ahead depend, to a great extent, on whether or not effective action will be taken to reverse some of the negative factors just cited. Here to stay are the expensive pollution control programs and the lighter, high-quality products, while break-in problems on the industry's new, complex facilities are largely a thing of the past and, for the most part, routine operations are now underway.

However, even though steelmakers have largely adapted to the rigors of high-technology production, it is most unlikely that significant increases in productivity will be realized unless the growth of volume becomes adequate to permit more efficient operating rates on a sustained basis. This makes it imperative that more effective measures be taken against steel imports which have been siphoning off the growth in domestic demand and that every effort be made to promote the use of steel over substitute materials.

Regarding the need to balance facilities, the 7% job development tax credit on new investments, should help steelmakers eliminate bottlenecks and round out capital projects, thereby enhancing the opportunity to productivity improvement.

Provided the all-important precondition of adequate volume is achieved, the prospects for increased productivity, based on the capabilities of the modern steelmaking equipment now in place, are extremely favorable. Further, despite the already advanced technological state of the industry, the surface has only been scratched in exploiting the vast potentials of such breakthroughs as continuous steel casting and direct reduction. It must be recognized, however, that steel's new technology has become increasingly difficult to install, operate, and maintain. Its successful implementation requires the fullest use of the talents of steel labor and management, and the newly established joint advisory committees on productivity could hold great promise toward this end.

Plant Productivity Committees

The Plant Productivity Committees were established by the agreement signed between the steelworkers and the industry in August 1971. Since that time several months have been given to their organization, and committees have actually been functioning in most plants throughout the nation. They consist of union and management representatives at the plant level.

In some plants encouraging progress has been made. The effort has resulted in a concerted attempt to improve product quality, to increase production, to curtail waste and cut down on absenteeism. All of these are directed toward improvement in man-hour output, and in a number of cases results already have been gratifying. On the other hand, there are some plants where virtually no progress has been made and few, if any, meetings have been held. However, it is still too early to pass judgment on the success or failure of the Productivity Committees. Presumably, they are to be permanent institutions and will outlast the present contract. Therefore, if most of them can function with effectiveness within the next year and continue to do so for years to come they will have achieved their purpose. Further, it is reasonable to expect that the success achieved in some plants will serve as proof to others that the concept of Productivity Committee is a workable one and beneficial to all. In addition to this, the United Steelworkers International officers have fully endorsed the committees as have many of the local officers. Thus, there is every reason to hope that the committees will be productive. Rather than pass a judgment at this time on their value and worth, it would be well to adopt a "wait and see" attitude and to offer whatever encouragement one possibly can to assure their successful operation.

In respect to productivity in 1971, it is interesting to note that in spite of the fact that raw steel production was 11 million tons below 1970 and shipments were almost 4 million tons below the previous year, there was an actual increase in productivity of 3.3%. Productivity, with 1967 as the base year equalling 100, rose to 104.8 in 1969, dropped sharply in 1970 to 101.7 and in 1971 recovered to 105.[a] Thus it was very slightly above the 1969 figure. Man-hours per ton of shipments were 11.7 in 1969, 11.9 in 1970, and 11.6 in 1971. If 1972 provides the increase in volume which is hoped for, there should be another productivity advance.

---

[a]Source: United States Department of Labor, Bureau of Labor Statistics, quoted on page 19 of the *1971 Annual Statistical Report* of the American Iron and Steel Institute.

5 Factors Affecting Steel Demand in the 1970s

The prospects for a higher volume of steel operations and improved productivity, which should in turn result in increased profits, are enhanced by the expectation of a growing steel market during the remainder of the 1970s. However, to participate in the enlarged market, steel companies must soon reach a decision on a new round of capital spending both to add new capacity and to eliminate bottlenecks now curtailing the output of recently installed, modern facilities. Individual decisions to spend must be tempered by current economic circumstances, which require the revision of traditional approaches to steel investment, and the amount of capital to be made available will largely be determined, as in the past, by the adjudged potential for future demand and, ultimately, future profits.

By 1980, assuming approximately the current relationship between steel imports and the domestic market, steel companies in the United States will have to produce 175 million net tons of raw steel to meet the requirements of both the private and public sectors of the economy. This projected annual rate of production, which is some 35 million net tons more than the industry's 1969 record output of 141.3 million net tons, is based on the expected growth of demand across a broad spectrum of traditional steel markets, most readily classified into the general categories of construction, transportation, durable goods, and containers. It takes into account potential new sources of steel demand just now starting to develop within traditional markets—sources typified by such new additions to the everyday industrial vocabulary as catalytic converters, Wankel engines, fast-breeder reactors, and electrostatic precipitators—and it also considers that a "second generation" steel demand has been developing as the nation's infrastructure has, in many respects, been growing woefully obsolete and is in need of extensive renewal.

The patterns of steel usage throughout the world provide a good indication of the stages of industrial development achieved by world economies, the most advanced being that of the United States, where per capita steel consumption since the mid-1960s has been roughly 60% higher than the average for ECSC countries. The growth of steel demand is generally slow until a certain point in industrialization and per capita income is reached. Thereafter, it increases rapidly until the nation's industrial machine has been built and the market for consumer durables becomes largely one of replacement. At this stage, long since achieved by the United States economy, additional income is directed to the purchase of services rather than durables, resulting in a more moderate growth in steel requirements or per capita steel consumption. Over the remainder of the decade, it is expected that steel consumption will grow at a faster rate than the nation's population, but not as fast as the gross national product (GNP), or in other terms, while per capita steel consumption will continue to rise, steel consumption per dollar of GNP will experience continued decline. However, while the growth of steel demand is expected to occur within this general framework, cognizance must be taken of the fact that the nation's development has, in many respects, come full circle, leaving an infrastructure that must be replaced and supplemented, and this will have a definite effect on the amount of steel required to satisfy the economy's needs in the latter part of the 1970s.

This chapter is not intended to be a tonnage forecast, product-by-product, of steel industry shipments; it is designed to point out the major factors and influences which will affect the growth and development of the steel-consuming industries for the remainder of the decade. Such an analysis, it is hoped, will thereby indicate the possibilities for growth in the steel industry itself.

Construction

The construction industry embraces a wide variety of activity, spanning both the private and public sectors of the economy. Both new construction and that for maintenance and repair are generally divided into two major categories–residential and nonresidential–which include various types of building activity originating from private and public sources. While the bulk of residential construction is privately financed (approximately 98% of the 14.4 million housing starts in the 1960s were private), much nonresidential construction activity is supported exclusively by public funding. Included in the private, nonresidential category are high-rise office buildings, other commercial and industrial buildings, public utility structures, institutional buildings, oil and gas pipelines, and a number of other less extensive types of construction. In the public sector, nonresidential construction includes schools, hospitals and government buildings; heavy construction such as highways, bridges, tunnels, and water and sewage treatment plants; and military installations and national defense facilities.

The construction industry is a very substantial consumer of steel, generally ranked second to the automotive industry in statistics reporting steel shipments. However, if the distribution of steel from the nation's service centers could be accurately measured, it is entirely possible that the construction industry would be the leading steel consumer, particularly when one considers that oil and gas pipelines fall under construction.

In 1971 the automotive industry purchased 17.5 million net tons of steel, or approximately 20% of total steel industry shipments. The construction industry purchased 14.5 million tons, or 16.6% of total steel industry shipments, including 4.9 million tons of steel in the form of "contractors' products" (i.e., steel consumed in such items as air conditioning systems, plumbing and central heating equipment, curtain wall panels, door hinges and fire escapes). In addition it is estimated that a substantial share of the 16.2 million tons shipped to the Steel Service Centers ultimately finds its way into construction. This is also true of the automotive industry, so it would be very difficult to determine in any given year which is the greater steel consumer. The significant fact is that the construction industry represents a very large market for steel.

During the remainder of the 1970s residential construction activity, spurred by an increased demand for housing and government subsidies, is expected to show significant growth. New housing starts should reach an annual rate of 2.4 million units toward the latter part of the decade, with total production for the 1970s aggregating over 22 million units. The production rate of 2.4 million units per year compares with an average annual rate for the 1960s of 1.6 million and a rate somewhat in excess of 2.0 million units for 1971. Earlier this year a peak rate of 2.6 million units prevailed for a time, and it may well be that in a random year or years between now and 1980 more than 2.4 million units will be built. The expected total of 22 million housing units will include homes built by conventional construction methods, prefabricated structures, modular or factory built housing, and mobile homes.

At the present time, mobile home construction represents the fastest growing segment of the housing market, and its continued growth is expected in the years ahead. According to the Mobile Home Manufacturers Association, a mobile home is a "transportable structure built on a chassis and designed to be used as a dwelling unit with or without a permanent foundation when connected with the required utilities." Unlike other types of housing construction, which are adaptable to single-family and multiple-family structures, the mobile home is limited to one-family needs and has proved particularly attractive to home buyers under thirty and over fifty-five years of age, people wanting minimal involvement with the problems of home ownership.

The recent popularity of mobile homes is indicated by the fact that they accounted for 50% of all the single-family dwellings built in 1971 and for 95% of all such dwellings priced at $15,000 and less. Their share of total housing unit production in 1971 was 33%, up sharply from 7.4% in 1960 and 21.6% in 1969. Sales in 1971 totaled approximately 500,000 units, valued at $3.5 billion, which breaks down to a median cost of about $7,000 vs. $25,000 for a conventionally built home. During the current year, mobile home output is expected to reach 580,000 to 600,000 units, with 700,000 units projected for 1973.

The sharp growth in mobile home output and sales indicates that the mobile home industry has been making significant progress in overcoming its long-time "trailer camp" image, a source of resistance to its product on the part of many communities. Top designers and architects have been lending their talents to the industry, and consequently many of today's mobile homes can compete on an esthetical basis with higher-priced traditional housing. An additional spur to demand has been the easing of government, FHA, and VA regulations covering loans to finance mobile home purchases.

For steelmakers, the bright outlook for mobile home production points to continued growth in what has already become a significant steel market. Depending on its size and the type of construction employed, a mobile home consumes 2300 to 2600 pounds of steel mill products. Its chassis frame uses 1500 to 1800 pounds of steel in the form of junior I-beams, U-beams, bar and rod stock, and tubing. Its roof and the nailer strips in its sidewalls require some 600 pounds of galvanized sheet and strip, and such parts as the unit's axle, springs, wheels and coupler assembly consume approximately 200 pounds of other steel products. Based on this year's mobile home output, which is expected to approach 600,000 units, the mobile home industry now constitutes an annual steel market approximating 750,000 net tons, not including the steel in such items as plumbing, heating, and ventilating equipment. In line with current estimates of mobile home output in the years ahead, this steel market should increase beyond 1 million net tons before 1975, with additional significant growth during the second half of the decade.

In the private, nonresidential construction category, industrial building has been sluggish owing principally to underutilized capacities combined with a greater than proportionate investment in machinery and equipment versus new plant structures. Business investment for expansion, the basis for new industrial construction, has been restricted as manufacturers have been channeling substantial capital into meeting the requirements of pollution control and occupational safety and health regulations. However, the installation of equipment necessary for compliance has already created a sizeable new demand for construction steels, one which should increase significantly as industrial building picks up in the years ahead.

Commercial building, another sluggish component of the private, nonresidential construction category, is being held down by an oversupply of office space in the nation's leading commercial cities, the result of a spurt in building after the mid-1960s. Vacancy rates have been high in New York, Los Angeles, Boston, Chicago, and Houston. High rise office buildings completed in New York in 1971 added 14.9 million sq. ft. of space, and another 23.2 million sq. ft. will be completed in the city this year. By 1973 New York's City Planning Commission expects a vacancy rate of 15% (34 million sq. ft. out of a total 225 million), with a normal vacancy rate of 4% looked for by 1976. In Los Angeles, 8.6 million sq. ft. of office space were completed in 1971, more than in New York during any year in the 1947-68 period. In Houston an oversupply of 20 million sq. ft. is a possibility by the end of 1973. While some builders feel that additional construction should be started now to accommodate an anticipated heavy demand for space by the mid-1970s, more are waiting for the current supply to be absorbed, postponing another upsurge in activity until middecade.

While office building construction has turned to the dull side, warehouse and shopping center construction has been extremely active and, significantly, has been providing a growth market for preengineered metal buildings. Sales of these structures, which are manufactured predominantly of steel, were somewhat in excess of $100 million in 1961, they increased to $400 million in 1969 and are expected to exceed $500 million by 1973. With a favorable outlook for demand, approximately 100 companies have entered the business, although 25 account for about three-fourths of the industry's sales volume. Kansas City-based Butler Manufacturing Company is the industry leader, with the Metal Products Division of Armco Steel rated the second largest producer.

The metal building is no longer the rusty old "toolshed" or the unsightly "Quonset hut," but provides an attractive means of construction for classrooms, gymnasiums, covered tennis courts, ice-skating rinks, motels, industrial plant buildings, the aforementioned warehouses and shopping centers, and a variety of structures on the farm. One major selling point that has attracted buyers of preengineered metal buildings is their impact on construction productivity. Using conventional means, a 40 ft. × 80 ft. farm structure, for example, could be built at a rate of 7-8 sq. ft. per man-hour, but a metal building of the same size can be raised at a rate of 40-50 sq. ft. per man-hour. In other terms, it can be completed in two days with a five man crew.

Another product designed to speed building construction, which should increase the demand for construction steels, is the light structural steel building panel. Individual panels are assembled in a plant, and when delivered to the construction site, they include the required door openings, windows, air conditioning, heating, ventilation and electrical outlets.

Continuing in the private, nonresidential construction sector, its oil and gas pipeline component has been a soft spot, which is reflected in the number of wells drilled. Another energy-related type of building, public utility structures, has been curtailed by delays over environmental issues. Meanwhile, a serious energy crisis has developed, making it likely that more effective measures will be taken to accelerate the development of our energy resources and the construction of additional power facilities. The Federal Power Commission recently predicted that the demand for electricity in the United States will quadruple by 1990 and estimated that this will require the construction of some 300 new power plants of more than 500 megawatts generating capacity, which comes to about one new power plant every twenty-five days. This portends a good steel market not only for construction but also for electrical machinery.

In the area of heavy construction, which covers highways, bridges, tunnels and water-sewage treatment plants, much remains to be done in the years ahead. Many of the bridges that were constructed in the early part of the century are woefully obsolete, for with the wider motor vehicles they can accommodate fewer lanes of traffic than were originally intended. In a number of instances, these connect areas within cities, such as Manhattan with other boroughs of New

York City, or the Triangle in Pittsburgh with the North and South Sides, making them a particular cause of traffic congestion during rush hours. In some instances these bridges have already been replaced, but many still remain.

The highway program which was adopted on a national basis several years ago is virtually complete as far as long stretches are concerned. However, a portion of the program is still to be finished, and this is predominately in the urban areas. Here, overpasses and cloverleafs must be installed as well as elevated highways, thus increasing the demand for heavy structural steel and plates. In one case in New York City, at the junction of several highways, a series of cloverleafs, and overpasses was constructed in an area of approximately two miles, requiring some 25,000 tons of steel. More of this type of construction is needed to reduce traffic congestion in urban areas, and this will be a far greater market for steel than long stretches of highway at ground level.

There has been considerable discussion about whether funds should be used for highways or mass transit, and undoubtedly there will be a compromise made so that both types of transportation will benefit. Mass transit, particularly if it involves a subway system, will require a substantial amount of structural steel and plates, so that the steel industry will benefit from either type of transportation increase, i.e., mass or individual.

The emphasis which has been placed on clean air and water during the last few years has given rise to considerable expenditure for sewage-treatment plants. Many of these have been installed and many are in the planning stage. During 1971, $1.8 billion was spent on sewer systems, a substantial portion of which was used for sewage-treatment plants. It is likely that this or even a larger amount will continue to be spent in the years ahead. Some of these systems are not particularly steel-intensive, but they do represent a considerable amount of steel, particularly in reinforced concrete structures.

Transportation

Transportation is a broad category covering both private and public, as well as mass and individual means of conveyance. Basically, it includes (1) motor vehicles, which, in turn, include passenger cars, trucks, and buses; (2) railroads, which handle both passenger and freight traffic, (3) shipping, which operates on the inland waterways, the Great Lakes, and high seas, and (4) air transport.

Among transportation equipment builders, motor vehicle producers have been the most prosperous, and represent the largest consuming market for steel. As previously indicated, in 1971, direct shipments of steel to this industry were 17.5 million tons. In addition there were unquestionably several millions of tons shipped indirectly from steel service centers, as well as significant tonnages imported from abroad. Further, it is interesting to note that two of the companies engaged in the production of motor vehicles have their own steel plants, namely, the Ford Motor Company, which has its steel division, and the International Harvester Company, which operates Wisconsin Steel Company, as a subsidiary.

The railroads are a basic necessity to the American economy and will continue as such, despite the fact that from time to time they have been unprofitable operations. The demand for steel for the railroads has been quite steady over a period of five years, averaging approximately 3.1 million tons per year. About one-third of this is in the form of rails, and the other two-thirds are used for the construction of freight cars, locomotives and other structures.

Shipbuilding has been a violently fluctuating cyclical industry which reached its high point during World War II. Since then it has been a relatively minor consumer of steel, averaging somewhat less than one million tons a year over the last ten years and consuming approximately 1% of steel mill shipments. However, a number of signs indicate that the industry will develop into a much stronger market for steel during the remainder of this decade.

The aircraft industry on a tonnage basis is a very small consumer of steel. For the last ten years it has averaged much less than 100,000 tons per year, or under one-tenth of 1% of steel industry shipments. However, it should be noted that a high percentage of the steel it consumes is either alloy or stainless steel. In 1968, the last year for which figures are available on alloy and stainless shipments to consumer industries, the aircraft industry took a total of 89,523 tons of all types of steels. Of this total, 59,353 tons were alloy steels and 11,703 tons were stainless steel. They represent, respectively, eight-tenths of 1% and 1.4% of these specialty steel shipments.

Automobiles

The automotive industry, the largest consumer of steel products since 1928, has been faced with serious import competition in the past few years.

Imports of automobiles have been increasing steadily since 1961, in which year a total of 279,000 passenger cars were imported. This was a considerable drop from the previous record year 1959 when 668,000 cars were imported. The reason for the decrease was the introduction by American producers of the compact car. However, 1962 witnessed an increase of almost 100,000 units imported, and from that point the number has risen every year. In 1965 it was 559,000, and in 1966, the second year in which the United States and Canada automotive products agreement functioned, imports jumped to 913,000 units. Since then they have risen to an all time record of 2.6 million units in 1971.

In addition to record passenger car imports in 1971, a new record of 239,000 units was established for truck imports. It should be noted that 135,000 of these came from Canada, and 99,000 from Japan.

In order to stem the tide of imports, the automobile producers in the United

States introduced the mini car in 1970. It was designed to compete with the small imported units with the hope that it would repeat the performance of the compacts which turned back imports in 1960 and 1961. Thus far the performance has been reasonably satisfactory, although imports for the first six months of 1972 were slightly above those for the same period in 1971. However, sales of imported cars have been down during the first six months as compared with the same period in the previous year. They amounted to 12% of the American market. This would indicate that a combination of the mini car and the new economic policy of the Nixon administration has been working in favor of the domestic automotive industry.

The policy when instituted in mid-August of 1971 was designed in part to stimulate the economy by boosting production in the focal automobile industry. The abolition of the 7% excise tax on automobiles, the establishment of a temporary 10% surcharge on imported cars, and the devaluation of the dollar, as well as the introduction and promotion of the mini car helped stem the loss of business to foreign manufacturers and contributed to increased sales of American cars.

In respect to imports, it should be noted that the 2.6 million figure for 1971 includes 800,000 cars produced in Canada by the subsidiaries of American automobile companies. Thus, if we exclude Canada, imports for 1971 amounted to just under 1.8 million units, which represented 15.3% of the domestic market.

Activity for the automotive industry during the remainder of the decade will depend in part on the success achieved in meeting the import challenge. Despite the fact that there has been a temporary set back in import sales, it can be reasonably expected that imports, particularly those manufactured by the leading foreign producers, will continue to be a major factor in the domestic market. The size and efficiency of their operations, their vast dealer net works, their technical competence and inventiveness all indicate that companies like Volkswagen and Toyota are not about to turn their backs and walk away from the most lucrative car market in the world. They will continue to press for a substantial participation in this market.

One highly significant factor in the entire automobile picture is the strict Federal regulations regarding automobile safety and pollution control which will go into effect by 1974 or 1975 at the very latest. Unless foreign producers are willing to invest substantial sums to conform with these regulations, there will certainly be a decline in imports. One possibility which could reverse this would be the adoption of similar regulations in other countries forcing their automobile industries to meet the same standards as those in the United States. This does not seem likely in the near future.

The formidable cost of compliance with American safety and pollution standards may well compel the leading exporters, such as Volkswagen and Toyota to restrict their shipments to the American market to perhaps one or two models. This could bring about a reduction in imports to a point where they might constitute less than 10% of the American market. Such a development would be a boon to the automobile industry and would increase steel shipments to it appreciably.

The introduction of the mini car has cut steel requirements to some extent. It is estimated that 1.2 tons are required for a mini car. This is opposed to 1.4, which was used in the original compacts, and 1.8 used generally in the larger cars. Offsetting this cut in tonnage is the fact that the automotive industry will add weight to the cars in the coming years to comply with ever increasing safety features. For example, in 1973 one of the compacts will increase in weight by 125 pounds virtually all of which will be steel. Medium size cars of one manufacturer will increase in weight up to 300 pounds, while others look to add 200 pounds on an average for all models. This is principally steel and will be a help to the steel industry. Experimental models have been made with extra bracing and extra beam supports throughout the body, as well as the heavier and more complex bumper structure, all of which will make the car heavier and make the imported cars less desirable from the point of view of safety since most of them will probably not develop the heavier bodies and frames. Thus the safety features, plus the pollution equipment on engines will aid the American automobile producers in retrieving a sizeable portion of the market which had been lost to imports.

Mention has already been made of the new market for stainless steel which is developing as a consequence of the automotive industry's efforts to control the air pollution from automobile engines. A 9 to 10 million passenger car year added to the production of 1.5 to 2 million trucks, and an emission control system for engines which will use as much as sixty-five to seventy pounds of stainless steel per vehicle could mean an additional market for over 400,000 tons of stainless steel annually. Whether such engines will be installed by foreign producers on most of their models is a question. Thus all of the aforementioned factors should have the effect of increased auto production during the rest of the decade with a consequent increase of steel shipments.

Railroads

The railroads, once the principal consumer of steel, have declined to a relatively minor position among the large steel consuming industries. In the past five years they have consumed an average of about 3.1 million tons annually which represents approximately 3.5% of total steel industry shipments. Rails and accessories represent about one-third of the steel, while two-thirds goes into freight cars, passenger cars, locomotives, and other uses.

Many of the railroads have been in deep trouble for the past few years as total net working capital for the industry declined from a high of $933.8 million in 1955 to a low of $56.1 million in 1969. There was a brief recovery in 1970 as the working capital figure rose to $109.1 million; however, 1971 witnessed a drop back to $84 million. In order to assist the railroads in alleviating their financial and other difficulties, the Federal government has allowed a number of mergers, the principal one being that between the Pennsylvania Railroad and the New York Central Railroad, which resulted in the largest system in the world. The merged unit subsequently went into bankruptcy.

The financial position of the railroads has resulted in a steady erosion of plant and equipment during the last decade. Rails have been bought in relatively small tonnages since 1958, and thus track is being replaced on the basis of 120-year life, at least 25% of the freight cars are 20 or more years old and many of the locomotives have become economically obsolete and are expensive to operate and maintain. Some notion of the rail replacement may be obtained from examining rail shipments during the last 15 years. The year 1957 was the last year in which more than a million tons of rails were shipped to the railroads; in that year the figure stood at 1.2 million tons. Subsequent to that there was a precipitous drop in shipments to 530,000 tons in 1958, and the all-time post-World War II low was reached in 1961 when shipments fell to 407,000 tons. Since then there has been a slight recovery, and in the past two years shipments have exceeded 800,000 tons. Nevertheless, the average for the fourteen-year period is less than 700,000 tons per year. When one considers that in the period from 1947 to 1957, the average was 1.7 million tons, the problem of rail replacement is brought into focus as a particularly acute one. It is true that some trackage has been abandoned during this period and in a number of instances a four-track section of the road has been reduced to a three-track section. In spite of this, however, the rail replacement remains a critical problem.

Another area in which the roads have deteriorated is the replacement of freight cars. The Association of American Railroads has called for a replacement of freight cars at the rate of 100,000 per year. However in the last fifteen years this figure has only been approached twice. In 1957 it stood at 99,502 and in 1966 at 88,899; 1971 represented an eight-year low as freight cars installed fell to 55,022. Almost the same situation applies to locomotives which have been installed in small numbers since 1965. In that year the figure was 1,387; it rose above this in 1966 to 1,419, but dropped to 978 in 1968 and was somewhat above the figure in 1971 with 1,190 units installed.

In spite of these statistics and the very poor showing of the railroads in the last decade, they are a most essential part of the American economy. It has been estimated that a railroad strike of one week would reduce the gross national product by over 5% and one of two months' duration would reduce GNP by about one-fourth and increase unemployment by approximately 30%. Consequently, something must be done to aid the railroads to achieve an economic comeback, so that they can function efficiently and safely as the nation's chief means of transportation.

There are a number of railroads which function profitably. However, there are six lines in bankruptcy and seventeen are considered marginal. As a result of this situation, there were several bills brought before Congress to help solve the railroads' problem. The principal one of these is the Hartke-Adams bill, or the Surface Transportation Act. If passed, it should help the railroads considerably. One of the principal provisions is to establish a government agency authorized to guarantee loans for all surface transportation up to a total of 5 billion dollars. Without question, the railroads would be the principal beneficiary of this section of the act. It would enable those in difficulty to borrow at relatively favorable rates since the government would guarantee repayment. Two other provisions of the bill could be particularly helpful to the railroads. One deals with rate changes, and the other with the abandonment of light density lines. In each instance the bill calls on the ICC to act within 120 days of the date on which notice is given. This could help appreciably since some ICC decisions have been held up for many months.

Some of the provisions in the original Surface Transportation Act have been shifted to other legislation. One of these, now part of the Highway Act of 1972, deals with railraod grade crossings. It would require the states to spend 5% of their highway trust funds on improving crossings and would allow this work to be done not only on Federal highways, which is currently the case, but also on secondary roads where it is badly needed. Another provision of the Surface Transportation Act, which has been shifted to the Revenue Act, forbids the states to tax transportation property at a higher rate than comparable property.

The year 1972 could represent a turning point for the railroads if the proper legislation is passed, and during the remainder of the decade they could be a much more efficient means of transportation. The system is essential to the economy of the nation, and it will have to replace rails at a much higher rate and also install locomotives and freight cars at a greater rate than has been done in the past decade. These measures will not only improve the railroads but will also make them a much stronger market for steel.

Shipbuilding

The shipbuilding industry consumed 1.2 million tons of steel products in 1971, which represented 1.3% of total steel industry shipments. The principal products were heavy structural shapes and plates. The former represented 155,000 tons, the latter 892,000 tons. The only other product of any significance was hot rolled bars, which totaled 41,000 tons. During the remainder of the decade there is every indication that shipbuilding will consume larger amounts of steel, perhaps two and one-half to three times as much as it did in 1971.

There are two basic reasons for optimism about this industry which has been a poor consumer of steel during most of the post-World War II period. The first is the passage of the Merchant Marine Act of 1970, which permitted subsidies from the Federal government to shipbuilding companies of up to 43% of the total cost of the vessel. In the fiscal year beginning July 1, 1972, the subsidies were reduced to 41%, which is still a substantial amount.

On June 30, 1972 some 1.9 million dead-weight tons of new shipping worth a total of 650 million dollars were contracted for. This tonnage was spread over sixteen ships, three of which were 265,000 ton tankers, the largest ever built in the United States. In addition, there were three other tankers rated at 225,000 tons, three 87,000 tankers and four small tankers rated at 35,000 tons, as well as three 18,900 ton cargo vessels. The contracts were divided among a number of shipyards, including the converted Brooklyn Navy Yard, Bethlehem Steel Corporation's Sparrows Point, Maryland yard, the National Steel and Shipbuilding Corporation yard in San Diego, the Todd Shipbuilding yards at San Pedro, California and the Bath Iron Works of Bath, Maine. The government's share of the construction cost amounted to 279 million dollars.

Another reason for optimism in shipbuilding growth is the close tie that shipping has with the energy crisis that the United States is facing today and will face in increasing intensity in the years ahead. It has been estimated that the BTUs required to fulfill our energy needs in 1985 will be at least double those used in 1970.

The question arises as to the energy source for these BTUs, and there are several possibilities: First, the United States has a tremendous reserve of coal which can be used in the conventional way or possibly gasified and used in that form. Secondly, there are the oil and natural gas reserves, a twofold energy source found separately or in combination, which currently supply most of our energy. However, the reserves of both oil and natural gas in the 48 contiguous states are dwindling. To a certain extent they have been bolstered by the discovery of 10 billion barrels of oil reserves in Alaska. With respect to these reserves there are problems of climate and contentions about the environment which may limit the production of Alaskan oil. Further, even if the reserves are fully utilized, the estimate of 10 billion barrels would be a relatively small increase for a nation consuming 5 billion barrels of oil per year. A third source of energy is atomic power; and there is a strong possibility that a large part of our requirement will be derived from this source, but there have been problems in meeting the timetables set for the development of atomic energy. Thus, there is a question as to how much of our needs it will meet in the next decade. A fourth source, water power, seems to offer little hope for expansion in the future.

The world reserves of oil are in excess of 600 billion barrels. Gas reserves on a world basis are also tremendous. However, the basic problem arises from the fact that oil and gas are found in nonindustrialized areas, such as the Middle East, and the industrialized areas—in many instances—must import either or both these fuels. Thus, the problem is one of transportation. If oil consumption in the United States by 1980 approaches the projections that have been made by the Department of the Interior, the National Petroleum Council, and the Maritime

Administration, imports will have to be simply tremendous. In 1970 the United States imported some 3 million barrels per day of all types of fuels. It is estimated conservatively that it will import 10.6 million barrels of oil per day by 1980 and 13.7 barrels per day by 1985. Other estimates place the 1985 daily imports as high as 18 million barrels. If imports reach more than 10 million barrels per day by 1980, this will constitute a total of approximately 3.5 billion barrels per year. The 250,000-ton deadweight tanker is capable of transporting about 1.7 million barrels of oil; thus it would take six of these large tankers per day to satisfy our import requirement, and if smaller tankers are used more would be required. Consequently, a very large fleet of jumbo size tankers will be required to fulfill our demands for energy by the end of the decade, if we use oil in the proportions indicated. Under the Maritime Act of 1970 presumably many of these tankers will be built in American shipyards. Thus the demand for steel in shipbuilding will be very substantial in the latter half of the 1970s.

In addition to oil it is more than possible that large amounts of liquified natural gas (LNG) will be imported. Such an operation will require a new type ship with highly specialized tanks. Present estimates of costs for a ship with 125,000 cubic meter capacity range from $85 million to $95 million per vessel. Those ships plying between American and foreign ports can be constructed in American shipyards with the aid of Federal subsidies. They will each require about 35,000 tons of steel, most of which will be structural and plates. However, one type of LNG carrier could use as much as 4,000 tons to 5,000 tons of specialty steel for the tank construction.

The number of LNG carriers which will be required quite naturally depends on the amount of this fuel to be transported. Predictions vary to some extent. The Federal Power Commission has stated that we may be importing as much as 2 trillion cubic feet by 1980. This would require a fairly large number of carriers, perhaps as many as eighty or ninety. At the present time the indications are that Federal subsidies would cover 25 to 27% of the cost of these vessels to make the shipyards in the United States competitive with those abroad. There are at least four shipyards in the United States which are equipped to build these carriers and during the next few years presumably there will be more. The LNG carrier, as well as the large supertanker will represent a substantial market for steel during the remainder of the decade.

The prospect for shipbuilding to carry oil and gas from distant points to the United States in huge quantities is a bright one. Depending upon how much of our energy comes from oil and gas rather than from coal and atomic power, these new fleets will vary in size. At worst they will be large, at best they will be enormous.

Durable Goods

Consumer durable goods consists principally of appliances, since automobiles are treated under transportation. The term "appliance" covers a large variety of items. Chief among these are: refrigerators, ranges (both gas and electric), clothes washers and driers, dishwashers, air-conditioners, food-disposal units, television sets, radios, phonographs, vacuum cleaners, and such minor appliances as toasters, irons, hotplates, frying pans and can openers.

There are relative degrees of market saturation for the major appliances; the most nearly complete instances are refrigerators, which are installed in 99.8% of all the wired homes in the nation, and black-and-white television sets which, likewise, are installed in 99.8% of the nation's wired homes. Other appliances with beyond 90% saturation of the market are clothes washers, vacuum cleaners, toasters, radios and coffeemakers. Those products that have a fairly low saturation rate are room air-conditioners, which have been installed in approximately 45% of the wired homes in the nation, dishwashers, which have reached a 30% saturation point, food disposal units, which are at 28% of the market, and electric waterheaters, which have a 34% saturation rate. Thus, in a number of instances, the market for appliances is a replacement market. However, there are new markets for major appliances when new families set up housekeeping. The rate of family formation depends to a great extent on the birthrate some 20 to 25 years back, and during the next few years, particularly by the end of the 1970s, it should be relatively high because of the birthrate in the middle to late 1950s.

Appliances have been a substantial consumer of steel for a number of years. This industry took over 2 million tons a year for the past 10 years, which represented approximately 2.5% of total steel mill shipments. The figure has been rather steady and has shown very little, if any, growth. During the same period, some of the major appliances have shown considerable growth. For example, in 1963 there were 4.1 million refrigerators shipped, and in 1971 this figure had risen to 5.7 million. Electric ranges in the same period rose from 1.9 million to 2.7 million. Dishwashers increased from 900,000 to 2.5 million. One of the greatest growth items was room air-conditioners, which advanced from 1.9 million in 1963 to 5.4 million in 1971, and, given the relatively low degree of market saturation, it seems that there are even greater sales increases in prospect for this particular appliance. One favorable fact in the future growth of appliance sales is the practice of installing new appliances in new homes, and if we consider that there should be more than 2 million housing starts per year, it is a market for approximately as many major appliances. This, in addition to replacement sales, should see the appliance market grow during the next 10 years.

The demand for producer durable goods, principally machinery, depends heavily on capital spending, which had little growth in 1971, as it rose only 1.9% above the previous year. This can be explained to a certain extent by the fact that there was considerable excess plant capacity available in many industries. In 1972, however, it looks as though the figure will advance by 9.7% over 1971 as spending will increase from $81.2 billion for 1971 to an estimated $89.1 billion in 1972.

Over the last five years steel industry shipments to the machinery and electrical machinery industries combined averaged approximately 8 million tons annually, or slightly under 9% of total steel shipments. The future of machinery as a market for steel depends on industrial growth and the need to replace economically obsolete plant and equipment. The presence of the 7% investment tax credit as well as the liberalization of depreciation regulations put in effect in 1971 will be helpful in maintaining this market for steel. However, it must be recognized that the market will be subject to cyclical fluctuations.

Metal Containers

Metal containers have constituted a substantial market for steel for several decades. In the last twenty years shipments to the container industry have shown a great stability with a slight growth rate during the last five years. In 1952 the industry took 5.6 million tons of steel products which constituted 8.1% of total steel industry shipments; in 1958 the 6.6 million tons taken by the container industry represented 11% of total steel industry shipments. This percentage was an all-time high. In subsequent years shipments hovered around 6.5 million tons on an average and represented between 7% and 8% of total industry shipments. In 1965 the tonnage figure topped 7 million for the first time, and with the exception of one year, namely 1966, has remained over 7 million tons ever since. The high point was reached in 1969 with 7.9 million tons shipped, in 1971 this figure dropped to 7.2 million.

The stability and relatively small tonnage growth over the years does not represent an adequate picture of steel performance in the container industry. In the early 1960s, in order to compete with aluminum and other substitutes, the steel industry substantially reduced the thickness of tin plate and sold it for a lower price. Before the advent of double-reduced tin plate, the average base box—which provides enough tin-plate to cover an area of 31,360 square inches—weighed ninety-two pounds. Double-reduced tin plate, which today represents 35 to 40% of tin mill shipments, weighs fifty-five pounds per base box. In effect, the lighter tin plate made it possible to make many more cans with the same tonnage. Therefore base boxes shipped, rather than tons shipped, indicates the true growth of tin plate as a packaging material. In 1962 there were 140.7 million base boxes shipped; this rose to 172.9 million in 1971.

Without question there has been very keen competition in the container field from a number of materials, including: aluminum, plastics, paper and varying combinations of these. Steel has stood up well against the competition and has made technological advances to sustain its position. One of the most notable has been the introduction within the past three years of tin-free steel, particularly for beverage containers. This product consists of a thin steel sheet coated with a flash of chromium in lieu of tin. Today it represents approximately 18% to 20% of "tin mill" shipments, and it looks as though it will continue to grow in the future. This is priced below conventional tin plate and gives the steel industry an additional competitive advantage.

One area in which the tin can or the can made from tin-free steel is expected to grow is in the beverage market. In 1969 there were 16.7 billion tin cans produced for beer, and this represented the first time that more than 50% of the beer packaged was in cans. In 1971 there were 18.4 billion beer cans made. In 1969 some 11.8 billion cans were made for soft drinks representing one-third of the total amount packaged; by 1971 the number had risen to 14.2 billion. This should increase to a point where it will represent 50% of the soft drinks packaged by 1973 or 1974. The aerosol container has been another growth market for the metal container. More than 2.5 billion of these were produced in 1969. This rose to 2.7 billion in 1971, and because more items are being packaged in these cans, which incidentally are larger cans, there is growth ahead. Tin plate should grow with the increase in population. However, it will be constantly harassed by other packaging materials and must continue to look for new developments.

One recent innovation in steel can manufacture that promises to help steelmakers recapture some of the tonnage taken by aluminum is Continental Can Company's shaping process for turning cylindrical can stock into specialty packaging, such as a 12 oz. miniature beer barrel. In the process, which is adaptable to both regular and aerosol packaging, steel cans with welded side seams are subjected to high-speed deforming, accomplished by means of internal pressure to force the walls of the can outward against a surrounding die with the desired shape.

Another recent advance in can manufacturing expected to benefit steelmakers is the "drawn-and-ironed" process for producing a two-piece steel can with the same neat appearance and light weight as the aluminum version, which is produced from a basically more expensive metal. The steel can has been a three-piece unit (combining a seamed body and two can ends); the more attractive two-piece aluminum can has been making significant inroads into its once-exclusive market (i.e., aluminum's penetration into the beer and soft-drink market is currently running at about 23 percent). Now, the drawn-and-ironed process developed for aluminum has been adapted to steel. Basically, it draws a can blank into a shallow cup, which is then elongated into a tall cylindrical container. Currently, the process requires the use of tin-plated can stock, the tin acting as a lubricant essential to the drawing operation. This limits the material cost advantage obtainable by using steel vs. aluminum. But once a suitable drawing lubricant is developed to permit the use of black plate, steel should make substantial inroads into the two-piece can market.

In addition to the influence of new can-making technology, the container market for steel in the years ahead will also be affected positively by environmental considerations. While the relatively high value of aluminum scrap has encouraged collection programs on the part of youth groups, civic organizations, etc., the recovery of discarded steel cans requires much less of a human resource input. The public's role in recycling steel cans simply involves not littering. Assuming that steel cans are properly disposed of, they can be separated magnetically from garbage and waste for recycling. This gives the steel can an edge from an environmental standpoint.

6 Conditions Determining the Investment Decision

The solution to the steel industry's problems, particularly those involving investment in additional facilities, demands a new approach, and this in part could consist of joint ventures and mergers. A new approach is required because the problems are new, and its necessity becomes evident when one considers the change in the world position of the United States steel industry and the support other governments give their steel industries.

In 1947, almost immediately after World War II when many of the industrial nations were prostrate, the United States produced 57% of the world's steel. As the other countries recovered and began to expand their steel industries the market share of the United States dropped continually. In 1950 it was 46%, in 1955 it was 39%, in 1960 it was 26%, in 1965 it was 26.2% and in 1971 it had fallen to 19.1%. Thus, instead of dominating the world steel industry America became a part of a much larger global industry.

The steel industry in other countries had grown rapidly during the 1950s, and during the 1960s it was to advance with even greater strides. The steel companies of Japan and Europe, often with the help of their governments, grew in size through internal expansion and mergers so that they were able to take advantage of the economies of scale in a way that was never before possible. Thus the advantage which the United States had, both prior to the war and for a decade and a half afterward, began to dwindle appreciably. An example of this is in the all-important light flat-rolled segment of the industry which was dominated for years by the United States. Before World War II there were very few large hot strip mills in existence outside of this country, principally because there were very few companies capable of producing enough steel to operate such units at capacity. With the growth of the industry abroad, this situation changed radically in the 1960s as a number of large hot strip mills were installed in Japan and Europe, and the United States became a net importer of sheets.

This was not only true of light flat-rolled products, but also applied to other basic items such as pipe, plates, and structurals. As a consequence, the decided competitive advantage which was maintained by the steel industry in the United States despite higher wages began to shrink and, in a number of cases, actually disappeared.

Foreign Mergers and Joint Ventures

The Japanese and European governments approved and in many cases encouraged mergers. The most spectacular instance of a steel merger since the formation of the United States Steel Corporation in 1901 was that of the Yawata and Fuji Steel Companies in Japan to form Nippon Steel Company. This new entity produced more steel in 1970, the first full year of its operation, than the United States Steel Corporation which, up to that time, had been the unchallenged leader in world steel production. U.S. Steel's output of crude steel in 1970 was 31.4 million net tons, while Nippon Steel's was 37 million net tons. The merger of these two Japanese giants in the steel industry would be akin to a merger of the United States Steel Corporation and Bethlehem Steel Corporation, if it were allowed. The formation of Nippon Steel was approved and encouraged by the Japanese government and was accomplished in spite of some minor opposition.

In Europe the steel industries in most countries have tried to realize the economies of scale, and in this effort their governments have encouraged mergers and joint ventures. In 1948 seven companies in France and Germany formed a joint venture. This firm, called Sollac, had steelmaking capacity of 3 million tons as well as an 80-inch, 10-stand hot strip mill. The basis for the joint venture was the fact that none of these companies in themselves had adequate steel to justify the installation of such a big mill. Since then, there have been two large mergers in France; one between Usinor and Lorraine-Escaut, with a resultant annual capacity of better than 6 million tons, and the other between the Wendel and Sidelor groups resulting in a 7.6 million metric ton company.

In Germany August Thyssen-Hutte AG is a combination of a number of companies. It acquired Phoenix Rheinrohr in 1965 and Oberhausen, a company with integrated steel mills, in 1968. Subsequently, an arrangement was made with Mannesmann AG by which all the pipe and tube plants of August Thyssen and Mannesmann were put together in a new company of which Mannesmann owned two-thirds and August Thyssen one-third. Thyssen absorbed the sheet rolling facilities of Mannesmann.

Another merger took place in Germany in 1966, between Hoesch and Dortmund. This combination resulted in a company with 5.5 million metric tons of raw steel. It was also tied in with the Dutch company Hoogovens, located at Ijmuiden, Holland. As far back as 1951, Hoogovens owned 43% of Dortmund. In 1964 and 1965 efforts were made to coordinate and combine steelmaking facilities in the Dortmund area. This resulted in the 1966 merger. As a consequence of the merger, the percentage ownership maintained by Hoogovens dropped to 14.5%. In 1966 Hoesch and Hoogovens agreed on a contract which would result in the distribution of labor and production in order to avoid duplication of investment. Talks have been held since that time between the respective boards of directors to consider questions of long-term planning in investment and the mutual usage of idle capacity, and a merger has been finalized.

In Belgium, with the encouragement of the government, Cockerill absorbed Esperance-Longdoz, a company one-third its size. Sidmar, another Belgium company with a plant located near Ghent, is in every sense of the word a joint venture whereby a number of companies participated in the construction of a 2.5 million metric ton grassroots plant. As it was organized Arbed owned 48%, Cockerill 29%, the Union Miniere 7%, the Falck Group in Italy 5%, the Schneider Group 3%, and other groups smaller percentages.

The nationalization of the British steel industry in 1967 created one company with 30 million net tons of steel by consolidating the fourteen largest raw steel producers in Great Britain. This venture was designed to improve the plight of British steel, however, it was politically rather than economically inspired. All of the companies in Great Britain with more than 475,000 tons of capacity were included in the nationalized company. It represented approximately 90% of the nation's carbon crude steel production; a much smaller percentage of specialty steels was represented primarily because the specialty steels were smaller companies with less than 475,000 tons of capacity.

In addition to the encouragement given mergers, steel industries outside of the United States have received significant government assistance, ranging from substantial capital loans at very low interest rates to nontariff protection against imports. Thus the policies which foreign governments have toward their steel industries have helped to rationalize their structures and make them far more efficient by avoiding the duplication of facilities. In Japan there is central planning in many instances. For example, a few years ago a decision was made on an industrywide basis to install seven blast furnaces; two were assigned to Yawata, two to Fuji, one to Kawasaki, one to Nippon Kokan, and one to Sumitomo. The rationale behind the allocation of blast furnaces was to control the amount of pig iron produced. It was thought that this would control investment throughout the industry, for pig iron was necessary to produce steel which would be rolled on the finishing facilities. At the present time, the Japanese steel companies, such as Kawasaki and Sumitomo do not feel that individually they could undertake the construction of a multiproduct, fully integrated, greenfield site plant. The current thinking is that such a venture would have to be undertaken jointly by two or more of these medium-size Japanese companies.

An example of rationalization in the Japanese steel industry was afforded during the last year through the organization of a depression cartel. It was necessary to curtail steel production because of the depression in Japan, and the function of the depression cartel was to apportion the reduction in output among the steel companies.

Because of antitrust policy in the United States, our steel industry has not had the same opportunity as its overseas counterparts to seek mergers, combinations, and joint ventures designed to make its facilities more efficient and productive and, at the same time, eliminate needless duplication. If an atmosphere can be created in the United States in which joint ventures and mergers among companies would be possible, provided in each individual case competition was maintained and customer interests protected, the members of the steel industry, and for that matter many other industries, would look for economically sound projects to be undertaken through joint ventures, or if a good economic fit presented itself, through mergers.

The steel industry in the United States has modern equipment second to none in the world. It also has efficient management and an effective labor force. However, it does lack the organization on a national basis that is found in Europe and Japan. Because of its structure, there is often needless duplication of expensive facilities, such as the ten hot strip mills installed during the 1960s.

The Decision to Employ New Technology

A second significant factor in the determination of steel industry investment is the decision on the part of management to employ new technology. It is affected by a number of items:

1. Management must be satisfied that the new technology, in whatever area it appears, can do the job, i.e., produce quality steel with the required chemical and physical specifications, and produce it economically
2. The new technology must be sufficiently superior in terms of its impact on operating costs and product quality to warrant the replacement of other facilities, particularly if they are in good working order. If the advantage is slight, and is outweighed by the investment cost involved, it would be extremely difficult for management to justify a change to a new method
3. To a certain extent individual company decisions to invest in new technology are governed by its actual or anticipated rate of introduction in the industry generally. If the industry's adoption of a new process is relatively slow, so that it has a minor impact on the total market, there will be less urgency to adopt it than there would be if there are many installations already in place which have a substantial effect on the market. If the latter is true, management will often be forced to make the decision in favor of installing new technology
4. Obviously a major, if not the most important determinant, is the financial position of the company
5. In the last two or three years, a new factor has come into consideration, and that is the effect that new technology will have on air and water pollution. If its use will cut down air and water pollution, this will be a strong favorable element in the decision to install it

Some examples will help to illustrate the above points. In respect to the first, there have been a number of questions raised during the postwar period about new technology and its ability to perform. Basically, there have been three fundamental breakthroughs, namely the basic oxygen steelmaking process, continuous casting, and direct reduction. The first of these, having long since overcome the problems involved in its conversion to large steel production, today is by far the dominant steelmaking process in virtually every major steel producing nation throughout the world, with the exception of the Soviet Union. Continuous casting which is revolutionary insofar as it bypasses several steps in the traditional steel making process, has had a long and agonizing breakin period. However, at the present time it seems that most of the difficulties connected with it have been solved, and it should yield good results. Direct reduction—an innovation which produces pellets with over 90% iron content bypassing coke ovens and blast furnaces—in many instances is still in the pilot stage, and there are only a few major plants in operation throughout the world. Problems connected with it must be solved before it can be looked on as a major producer of materials for the steel industry. In addition to these three breakthroughs, there have been other less dramatic but quite significant technological advances. These include high power electric furnaces and computerized and vastly improved rolling mills.

There were some questions in the minds of steel operators in the early and mid-1950s as to whether or not the basic oxygen process could make the full range of carbon steel products. Further, there was a matter of the size of the unit. In the beginning the units were approximately 35 tons per heat. Thus in the early 1950s when additional steel capacity was desperately needed in the United States, many felt that it would have been industrial folly to install a new type of technology about which there was some question. Thus open hearth furnaces were put in at that time because they could be counted on to do the job. Since then the improvements in the oxygen converter, often made by Americans, have increased the heat size to 300 tons and have made it capable of producing a full range of steels. As a consequence, it represents more than 85 million tons of this country's steel capacity in operation today.

There was some hesitation in the installation of continuous casting units because it was never proven that they could continually produce a good slab made of rimmed steel. Rimmed steel was a most important product and had many applications in the form of sheets. The problem was solved by developing other steels that could be continuously cast and had the same properties as rimmed steel.

Continuous casting units have been used since 1962 to produce other steels than rimmed steel. For example mini mills installed continuous casters to produce billets used for concrete reinforcing bar. However, for a number of years experiments with the continuous caster to produce a full line of steel products with consistent high quality did not meet with success. In fact, two companies who tied their entire fortunes to continuous casting met with disastrous results. One went into bankruptcy and the other showed heavy losses. In other instances where continuous casting was used, it was a supplement to conventional rolling and represented but a small portion of the company's output. Consequently, its mistakes and lack of performance were not too visible in overall figures. It seems, however, that most of the problems have been solved and the casters are now working up to potential. Thus we can look now for many more units to be installed in the near future.

In respect to the second point, is the new technology superior enough and reasonable enough in terms of installation costs to warrant replacing operating units that are in good order? Because of this consideration the installation of basic oxygen furnace capacity was delayed by some companies who had made substantial investments in the most modern open hearth equipment in the early 1950s. Basic oxygen showed a decided cost advantage over the old open hearths of as much as $8 a ton in some cases. In these cases replacements were made quickly. However, the new open hearths, particularly those equipped with oxygen lances, had such good cost performances that the difference between their operating costs and those of a basic oxygen converter were a few dollars a ton. Thus it was difficult and, in some cases where capital was short, impossible to justify the installation of the oxygen process to replace the recently constructed open hearths.

The third point is well illustrated by the installation of the second generation of continuous hot strip mills in the 1960s. Several companies who are significant producers of light flat rolled products made the decision to install new mills. These were so superior to their earlier counterparts both in terms of the quality of the product and the speed with which it could be produced that it was necessary for other producers to install comparable facilities if they were to remain competitive. By the mid 1960s at least five mills had either been installed or a decision had been made to install them. Consequently the other five were necessary to allow the other major sheet producers to compete for the market. Thus enough new technology in this area was put into operation to force the decision on the part of the remaining companies.

The drive in recent years to clean up the environment has required a considerable amount of capital expenditures throughout industry generally, and the steel industry will have to spend its share during the next few years. Both old and new equipment will receive considerable attention. Existing facilities that emit matter into the air and water will have to be equipped with devices to limit this as much as possible. New equipment will have to be designed to allow a minimum of air and water pollution.

Without doubt, the greatest problem faced by the steel industry is in the coke ovens, yet there are other facilities that present air pollution problems as well. They include sinter plants and the older open hearths that have not been equipped with precipitators. In the matter of the coke ovens, considerable effort has been put into research to solve the problem of the dust that arises from charging the coal into the ovens, the smoke and dust that is generated when the coke is actually pushed from the ovens, and the particulate matter which comes off when the coke is quenched. In respect to the first problem–charging the coal–there have been two approaches; one is the joint undertaking by the Environmental Protection Agency, and the American Iron and Steel Institute to develop a larry car which can charge coal and keep the dust down to an absolute minimum; the second is a new system developed by a combination of the Allied Chemical Corp. and Salem Corp. whereby coal is preheated and charged into the ovens through a pipe line, thus eliminating the dust which comes from the present method of charging. Both of these approaches have received close attention by the industry. It seems that the pipe line charging is more adaptable to a new set of coke ovens than it is to existing ovens. Thus the larry car development will probably be used on existing ovens, and in many instances the pipe line charging approach will be used on newly installed batteries of coke ovens that will replace those that have lived out their useful lives. The decision to place this new technology on coke ovens will be forced on the steel industry within the immediate future because of the pressure that it is under from environmental protection agencies. This is a capital expenditure which must be made in order for the industry to continue its operations.

Many of the existing open hearths are quite obsolete and not equipped with air pollution controls. They must either be so equipped in the very near future, or replaced by another steelmaking process. Since a number of them are small and inefficient by today's standards, it is highly unlikely that millions of dollars will be spent on this old, obsolete equipment to reduce air pollution. Whenever possible, the open hearth will be replaced by other steel making processes. Here the new Q-BOP will receive a great deal of attention, for unlike the existing basic oxygen process which blows the oxygen on top of the bath, the Q-BOP introduces the oxygen from the bottom of the vessel and the reaction is much less turbulent, thus reducing the amount of air pollution. This is certainly a point in favor of the new technology and will be a definite consideration in any decision on its installation.

Consequences of Noninvestment

If the steel industry does not invest in enough new equipment to provide the capacity needed to satisfy the demands of the economy in the late 1970s, the deficit in steel output will have to be made up by imports. This could mean that the import tonnage would increase substantially–perhaps to 30 million tons or even more. Under these circumstances, since foreign steel will not then be in competition with domestic steel but rather a supplement to it, import prices will rise substantially. They will be the equivalent of domestic prices or perhaps a bit higher. The foreign steel makers have not had a particularly profitable industry, and would welcome the opportunity to increase prices and improve their profit position. Thus, if the American companies do not provide the steel necessary, we will bring it in from abroad at higher prices and this will have a definite adverse impact on the balance of payments.

Another problem which can be foreseen if the domestic industry does not invest in plant and equipment in the next few years, would be the loss of business for the domestic mill builders. This could have a marked effect on the manufacturers of mill equipment and force them to reduce their capacity to produce steel mill facilities. Thus the skills and the capacity to build mills would be considerably curtailed, and perhaps even lost in one or two cases. The American Steel Industry could become dependent, at least in part, upon foreign mill builders for its equipment ten or fifteen years from now. Such a situation would not be particularly favorable either from the steel industry's point of view or from that of the balance-of-payments, since funds would have to be spent outside of the country to provide the necessary equipment for one of our basic domestic industries.

Thus the 1970s are critical years for the steel industry and those industries which supply it. The profit position of the steel industry in the next few years will be a determining factor in what happens in the area of investment, not only for the replacement of facilities, but for expansion which will be needed to supply the requirements of our industries by the end of the decade.

The attitude of the government, particularly the Department of Justice, will also have a strong influence on the ability of the steel industry to expand to meet the needs of the economy. If an atmosphere can be created in which it would be possible for joint ventures and mergers to take place when they are economically desirable and do not diminish competition, the industry will have an opportunity to restructure itself along sound economic lines. This will help give it the needed strength to serve the American economy and compete in the world market.

Notes

Chapter 2
Structural Changes and New Technology: Solutions to Steel's Investment Problem

1. Granite City Steel Company, Proxy Statement for Special Meeting of Stockholders, July 1971, p. 6.

2. Ibid., p. 4.

Chapter 3
Voluntary Restraint Agreements on Steel Imports

1. American Iron and Steel Institute, Communications and National Affairs Coordinating Committee, *Steel Imports–A National Concern* (Washington, D.C., American Iron and Steel Institute, July 1970) p. 1.

2. Letter of Agreement with Japan & Steel Exporters Association from Yoshihiro Inayama, Chairman, Japan Iron and Steel Exporters Association to The Honorable William P. Rogers, The Secretary of State, May 4, 1972.

Chapter 4
The Steel Labor Agreement and Steel Productivity

1. *Agreement between United States Steel Corporation and the United Steelworkers of America*, Pittsburgh, Pennsylvania, August 1, 1971, sections 1.3 to 1.6.

Index

Abel, I.W., 52, 75
Acme Steel Company:
  Interlake Iron Corporation, merger with:
    business prior to, 23
    plants prior to, 23
    terms of, 23
  Newport Steel Company, acquisition of, 22
  Riverdale facilities, constructed by, 22
  United States Steel Corporation, semi-finished obtained from, 22
Alan Wood Steel Company, labor negotiations, 1971, withdrawal from, 74
Allen, William M., 53
Allied Chemical Company, 119
Alloys Unlimited, Inc., Sharon Steel Corporation, bid for, 18
Aluminum cans:
  beverage market, penetration of, 110
  recycling of, 110-111
  replacement of, by steel, 3, 110
  substitutes for steel, as, 109
  two-piece container, 110
American Iron and Steel Institute:
  Consumers Union suit, named in, 71
  pollution control, coke ovens, and, 119
  Public Affairs Conference of, 52
American Export Industries, 26
American Magnesium Company, 14
American Metal Market, 55
Antitrust law:
  Census of Manufacturers, and, 21
  conglomerate movement, and, 20-21
  reinterpretation of, 20
A.O. Smith Company, 15
Apparent steel consumption, tonnages, 1957-1971, 49
Appliances (*see* Durable goods)
Arbed, 115
Argentina, 59
Armco/Boothe Corporation, 14
Armco Steel Corporation:
  Advanced Material Division of, 14
  Australian joint venture, and, 35
  direct reduction plant of, 38
  diversification by, 14-15
  HITCO, merger with, 14
  leasing business of:
    ARMCO/Boothe Corporation, 14
    GATX/Boothe Corporation, 14
  Metal Products Division of, 99
  non-USW plants, 74
  preengineered metal buildings, and, 99
  Reserve Mining Corporation, participation in, 34
  slab caster installed by, 42
  Strategic Planning Task Force of, 14
  super alloys business of, 14
  titanium business of, 14
A.R. Purdy Company, 13
Association of American Railroads, 104
Associazione Industrie Siderurgiche Italiane, 56, 68
Athlone Industries, Jessop Steel Company, take-over of, 18
Atomic energy, development of, 106-107
August Thyssen-Hutte AG, 35, 48, 114
Australia:
  iron ore in, 83
  joint venture possibility in, 35
  steel imports from, 1968-1971, 61
Austria, 61
Automobile industry:
  automobile excise tax, abolition of, 102
  automobile safety, 4, 102, 103
  import problem of, 101-103
  mini-car, 102, 103
  pollution control, and, 102, 103
  steel demand, and, 3, 100, 101, 103
Avnet, Inc., 17

Ball, George W., 53
Bank of America, 54
Basic oxygen steelmaking process:
  capacity in place, 117
  decisions to install, 118
  description of, 40
  flexibility of, 117
  furnace size, increase in, 117
  open hearth furnace:
    advantage over, 118
    replacement of, 39
  production time, reduction of, by, 40
Bath Iron Works, 106
Belgium-Luxembourg:
  output per man-hour in, 81
  steel imports from:
    1968-1971, 62
    decline in, 1969, 59
    increase in, 1971, 61
  steel output, 1955, 46
Bell & Howell Company, 54
Beneficiated iron ore, 83
Bessemer converter, 2
Bethlehem Steel Corporation:
  Bethlehem-Youngstown merger denial, 32
  diversification by, 13

Bethlehem Steel Corporation: *(cont.)*
Erie Mining Company, participation in, 34
labor bargaining team, 1971, 75
mergers, and, 114
shipments vs. import tonnage, 7, 51
shipyard, Sparrows Point, Maryland, 106
Blackie, William, 53
Boeing Company, 53
Borg-Warner Corporation, 54
Braniff Airways, 20
Brazil, 61
Bridge construction, 3, 99
British Independent Steel Producers Association, 68
British Steel Corporation, 68, 71, 115
Brooklyn Navy Yard, 106
Bruce-Gardner, Sir Douglas, 68
Brussels Tariff Nomenclature, 66, 68
Burke, Walter J., 75
Butler Manufacturing Company, 99

Canada, 58, 59, 79, 81, 101, 102
Capacity, steel:
additions to, 1960s, in, 2
future requirements for, 1, 3, 5
shortage of:
consequences of, 119-120
possibility of, 36
tonnage, current, 1
Capital investment, steel:
criteria for, new, 6ff.
during the 1960s, 2, 6
during the 1970s, 5
factors affecting, 116ff.
noninvestment, consequences of, 119-120
productivity, effect on, 1965-1970, 78
Catalytic converters, 95
Caterpillar Tractor Company, 53
Census of Manufacturers, antitrust policy, and, 21
Central planning, Japanese steel industry, in, 115
Cerro Corporation, 13
CF&I Steel Corporation:
Crane Company, take-over by, 18
labor negotiations, 1971, withdrawal from, 74
take-over bid for, 17, 18
Chambre Syndicale de la Siderurgie Francaise, 57, 67
Chas. Pfizer & Co. Inc., 54
Chase Manhattan Bank, 54
Cleveland Cliffs Iron Company, 26, 27
Cliffs-St. Clair Corporation, 27
Cockerill, 114, 115
Cold finished steel bar imports, voluntary limits on, 66, 69
Colt Industries, Crucible Steel Corporation, take-over of, 17
Columbia Tool Steel Company, Teledyne, Inc., take-over by, 18
Comite de la Siderurgie Belge, 67
Committee on Ways and Means, 52, 54, 57
Conglomerate movement:
antitrust law, and, 20-21
steel take-overs, and, 16ff.
Construction industry:
building panels, and, 99
commercial building, 98-99
heavy construction, 99
housing starts, 1970s, 97
mobile home segment of, 97-98
national highway program, and, 100
nonresidential activity, 98-100
office building construction, 98
oil and gas pipeline construction, 99
power plant construction, 99
preengineered metal buildings, and, 99
private vs. public activity, 96
residential activity, 97-98
shopping center construction, 99
steel demand, and, 96-97
sub-divisions of, 96
Consumer Price Index, 76
Consumers Union of the United States, Inc., 71
Consumption, steel (*see* Apparent steel consumption)
Containers (*see* Metal containers)
Continental Can Company, 110
Continental Steel Corporation, Penn-Dixie Cement Corporation, take-over by, 18
Continuous casting, steel:
commitments to install, 6-7
facilities replaced by, 42
joint ventures, and, 43
losses resulting from, 117-118
mini-mills, and, 117
prospects for, 42
rimmed steel, and, 117
slab casting installations, 42
steel productivity, and, 90
wide slabs, difficulty with, 42
yields, steel, and, 3, 43
Cost-of-living adjustment, steel, 1971, 76
Costs, steel industry:
labor, for, 7, 9, 75-76
operating costs, 8
plant and equipment, for, 2, 7, 85, 86
raw materials, for, 7
technology and reduction of, 36ff.

Council of Economic Advisers, 73, 74
Crane Company, CF&I Steel Corporation, take-over of, 18
Crucible Steel Corporation:
  Colt Industries, subsidiary of, 17
  take-over bid for, 11, 17-18
  Walter Kidde & Co. Inc., approached by, 17
  Whittaker Corporation, approached by, 17
Cyclops Corporation:
  Detroit Steel Corporation, merger with:
    plants acquired, 27
    stock purchases to acquire, 26-27
  Empire-Reeves Steel Division of, 26
  labor contract extension, 1971, 74
  Sawhill Tubular Division of, 26
  Sharon Steel Corporation, merger attempted with, 18, 26
  Universal Cyclops Speciality Steel Division of, 26
Czechoslovakia, 59

Dean, R. Hal, 53
Deere & Company, 54
Demand, steel:
  aircraft industry, and, 101
  appliances, and, 108
  automobile industry:
    growth in 1970s, 3
    pollution control, and, 103
    principal steel consumer, as, 101
    safety, and, 103
    shipments to, 1971, 100
  can-making technology, and, 110
  construction industry vs. automobile industry, 96-97
  construction markets in, 96-100
  double-reduced tinplate, effect on, 109
  durable goods, and, 107-109
  freight cars, and, 104
  gross national product, relation to, 96
  growth in, 1, 3-5
  heavy construction, and, 99
  housing starts, and, 97
  industrial development, and, 95
  LNG carriers, and, 107
  metal containers, and, 109-111
  mobile homes, and, 97-98
  national highway program, and, 100
  office building construction, and, 98
  oil and gas pipelines, and, 99
  oil tanker construction, and, 107
  pollution control, and, 100
  preengineered metal buildings, and, 99
  private, non-residential construction, and, 98
  private sector, in, 3
  public sector, in, 3
  railroads, and, 101, 103, 104, 105
  "second generation" of, 95
  shipbuilding industry, and, 101, 105, 107
  tin-free steel, and, 109-110
Depression cartel, Japanese steel industry, in, 115
Detroit Steel Corporation:
  Cyclops Incorporated, merger with, 26-27
  General Motors strike, 1970, impact on, 27
  markets, steel, of, 27
  plant locations of, 27
  Portsmouth, Ohio plant of, 26
  profit performance, 1969-1970, 27
Direct reduction:
  definition of, 37
  facilities eliminated by, 37
  HYL Process for, 38
  interest in, reasons for, 37
  investment cost reduction, and, 37, 38
  operating plants, 38
  prospects for, 38
  SLRN Process for, 38
Diversification, steel companies, by, 2, 11ff.
D'onofrio, Mario, 68
Dortmund, 114
Durable goods:
  appliances:
    market saturation for, 108
    types of, 107-108
  family formations, and, 108
  housing starts, and, 108
  steel consumption category, as, 107-108
  steel demand, and:
    consumer durables, 108
    producer durables, 108-109
Duvall, Donald C., 25

Electric energy:
  requirements for, 3
  sources of, 106
Electric furnace steelmaking process:
  cost advantages of, 41
  direct reduction, and, 37
  furnace size, increase in, 41
  high power units, 117
  increased output of, 1, 39, 41
Electricity demand, 1990, by, 99
Electrostatic precipitators, 95
Environmental Protection Agency, 4, 119
Erie Mining Company, 23, 34
Esperance-Longdoz, 114

European Coal and Steel Community, 46, 47, 55, 57, 58, 59, 65, 66, 67, 95
European steel industry:
business decline, 1971, and, 60
cold reduction mills installed, 47, 48
cold-rolled sheet imports from, 1970-1971, 47
expansion, by early 1960s, 46
hot strip mills installed, 47, 48
mergers in:
companies participating, 48, 113ff.
government encouragement of, 113
productivity comparisons, 87
rationalization of investment in, 9
steel imports from, 1969, 58, 59
voluntary quota tonnage adopted, 1968, 55
voluntary restraint agreement, 1972, and, 62ff.
Exports, steel, United States:
dollar value of, 1971, 61
tonnages, 1957-1971, 49

Fabricated structural steel imports, voluntary limitations on, 66, 69
Falck Group, 115
Fast-breeder reactors, 95
Federal Housing Administration, 98
Federal Power Commission, 99, 107
Ferry, Jaques, 57, 67
Finishing facilities, steel, 5, 7, 86, 118
First National City Bank, 54
Firth Sterling, Inc., Teledyne, Inc., takeover by, 18
Ford, Henry II, 53
Ford Motor Company, 53, 101
France, 46
Fried. Krupp Huttenwerke AG, 71
Fuji Iron and Steel Company, Ltd., 48, 56, 114, 115

Gas production, 3
Gas reserves, 3, 106
GATX/Boothe Corporation, 14
General Agreement on Tariffs and Trade, 53
General American Transportation Corporation, 13
General Motors Corporation:
labor settlement, 1970, 73
opposition to steel import quotas, 54
strike at, 1970:
Detroit Steel Corporation, impact on, 27
steel negotiations, and, 73
Georgetown Steel Company, 38
Germany (*see* West German steel industry)
Goldman Sachs and Company, 25
Grace, J. Peter, 53
Granite City Steel Company:
hot rolled coil purchases, 29
market area served by, 30
National Steel Corporation, merger with:
agreement and terms of, 30
benefits of, 30-32
properties and facilities, 1970, 29
ranking, steel industry, in, 1970, 28
Greenfield site steel plants:
joint ventures, and, 33
prospects for construction of, 33
Groupment des Hauts Fourneaux et Acieries Belges, 5
Groupment des Industries Siderurgiques Luxembourgeoises, 57, 68

Haggerty, Patrick E., 54
Hanna Mining Company, 29
Hartke-Adams bill, 105
Hartke, Vance, Senator, 52
Hastings Aluminum Products, 14
Heinz, H.J. II, 54
Hewitt, William A., 54
Hewlett-Packard Company, 54
Highway Act of 1972, 105
HITCO, 14
H.J. Heinz Company, 54
Hoesch, 114
Hojalata y Lamina Steel Company, 38
Hoogovens, 114
Hospital construction, 3
Hot strip mills:
capacity of, 8
cost of, 7
decisions to invest in, 118
installations, additional, prospects for, 5
investment in, 1960s, 2
Japan and Europe, in, 46, 113
"second generation," 7, 118
Housing starts:
durable goods demand, and, 108
outlook, 1970s, 97
Houthakker, Hendrik, 74
Howe Engineering Company, 14
Hunt Foods, 23, 25
HYL direct reduction process, 38

IBM World Trade Corporation, 54
Imports, steel:
apparent consumption, percent of, 49
Bethlehem Steel Corporation shipments, compared to, 51
geographic limits on, 64, 66, 69
geographic markets, and, 49-50

job losses, and, 51
labor contract negotiations, 1971, and, 73
light-flat products, of, 7
low prices, and, 47
market penetration of, 3, 9, 49
nonquota countries, from, 1968-1971, 61
product categories, by, 47, 63, 113
quota legislation, and, 52
specialty steels, of, 59-60
steel productivity, and, 88
steel strike, 1959, and, 45, 47
tonnages, 1957-1971, 7, 45-46
tool steels, of, 60
trade deficit resulting from, 1971, 60-61
trans-Alaska pipeline, and, 50-51, 60
United States Steel Corporation shipments, compared to, 51
voluntary restraint agreements, and, 51, 54ff.
wire rods, prices of, 59
world output, growth of, and, 46
Inayama, Yoshihiro, 55, 70
India, 58, 59
Ingersoll, Robert S., 54
Inland Steel Company:
A.O. Smith Company, joint venture with, 15
diversification of, 15
Inland Steel Development Corporation, 15
Inland Steel Urban Development Corporation, 15
On-Line Decisions, Inc., interest in, 15
Scholz Homes, Inc., acquisition of, 15
slab caster installed by, 42
Interest payments, steel companies, by, 8
Interlake Iron Corporation:
Acme Steel Company, merger with, 21-23
business prior to merger, 23
plants prior to merger, 23
Interlake Steel Corporation:
Erie Mining Company, participation in, 34
formation of, 23
International Harvester Company, 101
International Packers, Ltd., 54
Invested capital, steel, in, 8
Investment, steel industry (*see* Capital investment, steel)
Iron Ore Company of Canada, 34
Italy, 46, 61, 81

Japan Iron & Steel Exporters' Association, 55, 68, 70, 71
Japanese steel industry:
business decline, 1971, and, 60
central planning in, 115
cold reduction mills installed, 47, 48
cold-rolled sheet imports from, 1970-1971, 47
contracted labor in, 87
depression cartel in, 115
expansion, early 1960s, to, 46
expansion plans, reduction of, 35
hot strip mills installed, 47, 48
imports from, 1968-1971, 58, 61, 62
investment, United States, influence on, 7
mergers in, 48, 113ff.
output of:
1947, 1955, 46
post-war limits on, 46
output per man-hour in, 79-81
productivity of vs. United States steel industry, 87
rationalization of investment by, 9, 115
tinplate exports to Hawaii, 50
trans-Alaska pipeline, and, 50-51
voluntary quota proposed by, 54
voluntary quota tonnage adopted, 1968, 55
voluntary restraint agreement, 1972, and, 62ff.
Jenkins, Judge, 32
Jessop Steel Company, Athlone, Industries, take-over by, 18
Johnson, Lyndon B., President, 54
Joint ventures, steel:
antitrust policy, United States, and, 115
Australia, possibility in, 35
capacity requirements, and, 33
capital costs, and, 33
capital spending, effect on, 9, 11
duplicated investment, role in eliminating, 34
economies of scale, and, 114
encouragement by foreign governments, 114-115
finished steel products, for, 34
foreign, 113-116
greenfield site plants, and, 33
iron ore developments, in, 34
Justice Department, and, 35
New Orleans, advantage of, at, 35
semifinished steel products, for, 34
technology, new steel, and, 36
Jones & Laughlin Steel Corporation, Ling-Temco-Vought, acquisition by, 17-20
Justice Department:
Cerro Corporation, acquisition of, Beth-

Justice Department: *(cont.)*
lehem Steel Corporation, by, adverse ruling on, 13
conglomerates, regulation of, by, 19-20
joint ventures, and, 35
Jones & Laughlin Steel Corporation, acquisition of, Ling-Temco-Vought, by, suit against, 20
National Steel Corporation–Granite City Steel Company merger, and, 28, 30
Sharon Steel Corporation–Cyclops Corporation merger attempt, opposition to, 18, 26
steel expansion, 1970s, influence on, 120

Kaiser Steel Corporation:
Australian joint venture, and, 35
labor contract expiration, 1971, 74
Katz, Julius L., 71
Kawasaki Steel Corporation, 56, 115
Kennedy administration, steel price rollback, and, 16
Kennedy Round Agreements, 53
Kettering family, 25
Kobe Steel Works, Ltd., 56
Korean Republic, 59
Korhumel Steel & Aluminum Company, 13
Kusan Corporation, 13

Labor agreement, steel, 1971:
cost of living adjustment, 76
grievance procedures, 77
holiday granted, 77
Nixon administration, and, 74
pattern followed in, 76
productivity, emphasis on, 77-78
productivity committees, 76, 78
rates of pay, 77-78
wage-price freeze, and, 76
Labor negotiations, steel, 1971:
can industry settlement, and, 75
capacity, steel, percent not affected by, 75
companies withdrawing from, 74
contract extensions, and, 74
incentives, and, 75
local issues, and, 75
mini-mills, and, 75
strike, possibility of, and, 75
tentative agreement reached in, 75
Larry, R. Heath, 75
Lehman Brothers International, 53
Levy, Gustave L., 25
Liquified natural gas (LNG), 4, 107
Linen, James A., 54
Ling-Temco-Vought:
Jones & Laughlin Steel Corporation:
acquisition of, events following, 19-20
cash offer for, 18
Youngstown Sheet & Tube Company, merger talks with, 17
Littman, Mark, 68
Litton Industries, Inc., 54
LNG tankers:
capacity of, 4
construction of, 107
cost of, 4
requirements for, 4
steel in, 4
terminal facilities for, 4
Lone Star Steel Company, Northwest Industries, take-over by, 18
Long-term debt, steel in, 8
Lorraine-Escaut, 114
Lukens Steel Company:
labor contract extension, 1971, 74
slab caster installed by, 42
Lykes Corporation, Youngstown Sheet & Tube Company, merger with, 17, 18

Mahony, Joseph P., 75
Mannesmann, A.G., 114
Manuelli, Ernesto, 56
Maritime Act of 1970, 107
Maritime Administration, 106-107
Maxwell, Allison R., Jr., 25
McLouth Steel Corporation:
modernization, steel facilities, of, 1, 2
slab caster installed by, 42
Melchett, Lord, 68
Merchant Marine Act of 1970, 4, 105
Mergers, steel:
antitrust policy, United States, and, 115
benefits derived from, 36
capital spending, effect on, 11
competition, effect on, 36
Cyclops Incorporated–Detroit Steel Corporation, 26-27
economies of scale, and, 114
encouragement of, foreign governments, by, 114-115
Europe, in, 48
foreign, 113-116
Interlake Iron Corporation–Acme Steel Company, 21-23
intraindustry, 11
Japan, in, 48
National Steel Corporation–Granite City Steel Company, 28-32
overseas investment, and, 48
Wheeling Steel Corporation–Pittsburgh

Steel Company, 23-25
Mesaba Cliffs Mining Company, 27
Mesabi Range, 83
Metal containers:
aluminum vs. steel in, 109-110
beverage market, and, 110
double-reduced tinplate in, 109
"drawn-and-ironed" process for, 110
growth prospects, 1970s, 3
recycling of, 110-111
shaping process for, 110
steel demand, and, 109-111
steel substitutes, use of, in, 109
tin-free steel in, 109-110
Mexico, 38, 50, 58
Midland-Ross Corporation, 38
Mills, Wilbur D., 58
Mini car:
introduction of, 102
steel requirements for, 103
Mini-mills, steel:
continuous casting, and, 117
steel industry structure, effect on, 11
Mobile Home Manufacturers Association, 97
Mobile homes, 3, 96-98
Monsanto Chemical Company, 24
Moore, George S., 54
Morris, Robert M., 24, 25
Multicon, 13

Nakayamma Steel Works, Ltd., 56
National Can Corporation, labor settlement, 1971, 73-74
National highway program, 100
National Petroleum Council, 106
National Steel and Shipbuilding Corporation, 106
National Steel Corporation:
aluminum operations of, 13-14
diversification by, 13-14
Granite City Steel Company, merger with:
agreement and terms of, 30
benefits of, 30-32
competition, effect on, 30
Justice Department, and, 28, 30
properties and facilities involved, 28-29
reasons for, 29-30
magnesium operations of, 14
market area served by, 30
non-USW plants, 74
properties and facilities, 1970, 28-29
ranking, steel industry, in, 1970, 28
slab caster installed by, 42
steel service centers of, 13
Nationalization, British steel industry, 115
Netherlands, 62, 81
New York Central Railroad, 104
New York City:
highway construction in, 100
Planning Commission of, 98
Newport Steel Company, 22
Nippon Kokan Kabushiki Kaisha, 56, 115
Nippon Steel Corporation, 71, 114
Nisshin Steel Co., Ltd., 56
Nixon administration:
new economic policy of, 102
surcharge on imports, 60
wage-price policy of, 78, 85
Nixon, Richard M., President, 62, 70, 74
Northwest Industries, Lone Star Steel Company, take-over of, 18
Northwestern Steel and Wire Company, 41
Nova Scotia, 41
NVF Company, Sharon Steel Corporation, take-over of, 17-19

Oberhausen, 114
Occupational safety and health regulations, 98
O'Connell, John J., 75
Office building construction, 98
Oil and gas pipelines, 99
Oil reserves, world, 106
Oil tanker construction, 107
Okonite Company, 20
On-Line Decisions, Inc., 15
Open hearth steelmaking process, 2, 3, 39, 118
Oregon Steel Mills, 38
Osaka Iron & Steel Co., Ltd., 56
Output per man-hour:
Bureau of Labor Statistics:
qualifications to, 86-87
use by, 85-87
international comparisons of, 87-88
Japanese steel industry, in, 87
lag in steel, 78-81
partial productivity ratio, as, 85
petroleum refining, in, 79
Phase 2 stabilization, and, 78, 85
productivity measurement, as, 78, 85-87
quality improvements, and, 85
selected countries, in, 79-81
selected manufacturing industries, in, 79-80
shortcomings of, 85-87
West German steel industry, in, 87

Packard, David, 54
Patton, Thomas F., 52
Penn-Central Railroad, 104

Penn-Dixie Cement Corporation, Continental Steel Corporation, take-over of, 18
Pennsylvania Railroad, 104
Peterson, Peter G., 54
Peterson, Rudolph A., 54
Petroleum refining, output per man-hours, in, 79
Phase 2 stabilization program, 78, 85
Philadelphia Reading Railroad, 18
Philippines, 61
Phoenix Rheinrohr, 114
Phoenix Steel Corporation, slab caster installed by, 42
Pittsburgh Chemical Company, 12
Pittsburgh Steel Company:
  Wheeling Steel Corporation, merger with, 23-25
  Wheeling stock, purchase of, 1967, 25
Poland, 61
Pollution control:
  automobile, in, 4, 102, 103
  business investment, effect on, 98
  capital requirements, steel, and, 2, 9
  coke ovens, and, 43, 119
  cost of, 4, 9, 43-44
  investment in, 43
  investment tax credits, and, 44
  open hearth furnaces, and, 39
  Q-BOP steelmaking process, in, 40
  stainless steel demand, and, 4
  steel demand, and, 4, 100
Population, United States, by 1980, 3
Power plant construction, 99
Powers, John J., Jr., 54
Preengineered metal buildings, 99
Price-earnings ratios, steel stocks, for, 16-17
Production, steel:
  raw steel, of, 1
  United States as percent of world, 113
Productivity committees, steel:
  formation of, 76-77
  progress of, 92-93
Productivity, steel:
  capital costs, and, 91
  concept of, 81-84
  continuous casting, and, 90
  facilities, complexity of, and, 89-90
  facilities imbalance, and, 91
  incentives, new equipment, on, and, 91
  increases in:
    1971, 93
    prospects for, 91-93
  input categories in:
    machinery, 83-84, 85
    manpower, 82, 85
    materials, 82-83, 85
  international comparisons of, 87-88
  Japan, in, 87
  labor market conditions, and, 91
  light-weight steels, and, 84, 88-89
  multiple product lines, and, 84
  output per man-hour measure of, 78, 85-87
  Phase 2 stabilization, and, 78, 85
  pollution control, and, 91
  quality improvements, and, 89
  rolling mill, at, 86
  steel imports, and, 88
  steel industry, lag in, reasons for, 88-91
  steel volume, and, 88, 95
  tax credit, role in, 92
  "thin" tinplate, and, 89
  West Germany, in, 87
Profits, steel, 1, 5, 6, 8, 9
PTO-pricing, steel, 24

Q-BOP steelmaking process:
  cost advantages of, 40
  hot metal yield, and, 40
  open hearth, replacement for, 119
  pollution control, and, 40
  United States Steel Corporation, installation by, 41
Quonset hut, 99
Quota legislation, steel imports:
  opposition to, 52-54
  support for, 52

Railroads:
  financial position of, 103-104
  freight car replacement, 104
  legislation, pending, and, 105
  mergers of, 104
  nation's economy, and, 104
  steel demand, and, 101, 103, 104, 105
  track replacement program, 104
Ralston Purina Company, 53
Rationalization, steel investment, of:
  domestic steel companies, among, 9
  overseas steel companies, among, 7
Raw materials, steelmaking, for, 1, 82-83
Raw steel requirements, 1980, 95
Reorganization, steel industry, 11ff.
Republic Foil Company, 14
Republic of South Africa, 58, 59
Republic Steel Corporation:
  diversification of, 15-16
  labor bargaining team, 1971, 75
  real estate interests, 15-16
  Republic Steel Enterprises, Inc., subsidiary, 15
  Reserve Mining Corporation, partici-

pation in, 34
sales of vs. National Steel Corporation, 28
slab caster installed by, 42
Reserve Mining Corporation, 34
Revenue Act of 1962, 2
Revenue Act of 1972, 105
Roanoke Electric Steel Corporation, 42
Roche, James M., 54
Rockefeller, David, 54
Rogers, William P., 71
Rosenberg, Louis, Federal Judge, 20
Rusk, Dean, 58

Salem Corporation, 119
Samuels, Nathaniel, 70
Schmit, Rene, 57, 68
Schneider Group, 115
Scholz Homes, Inc., 115
School construction, 3
Senate Committee on Finance, 52, 54, 57
Sendler, Hans-Jorg, 67
Shaped can, 3
Sharon Steel Corporation:
Alloys Unlimited, Inc., approach by, 18
Cyclops Corporation, merger attempted with, 18, 26
NVF Company, take-over by, 17-19
Shaver, J. Warren, 75
Shipbuilding:
contracted tonnage, mid-1972, 106
energy crisis, and, 106-107
LNG tanker construction, 107
Maritime Act of 1970, and, 107
Merchant Marine Act of 1970, and 107
oil tanker construction, 107
optimism about, reasons for, 105-106
prospects for, 107
steel demand, and, 101, 105, 107
Shipments, steel:
light-flat products, of, 8
tonnages, 1957-1971, 49
Shopping center construction, 99
Sidelor group, 114
Sidmar, 114
Signal Corporation, 17
Simon, Norton, 23, 24, 25
SIP steelmaking process, 41
SLRN direct reduction process, 38
Sohl, Hans-Gunther, 57
Sollac, 114
South Wire Company, 14
Spain, 35, 59
Specialty steel imports:
tonnage of, 59-60
voluntary limits on, 64, 65-66, 68
Speer, Edgar, 13
Stainless steel, automobile market for, 4
Steel Company of Canada, Erie Mining Company, participation in, 34
Steel service centers, 96, 97, 100
Stockholders' equity, steel:
increases in, 8
returns on, 5
Stock market:
conglomerate take-overs, steel, and, 16
steel equities, performance of, 16
top eight steel companies, performance of, 17
Structural change, steel industry, 11ff.
Sumitomo Metal Industries, Ltd., 56, 115
Surface Transportation Act, 105
Sweden, 61, 81
Switzerland, 81
Sydney Steel Corporation, 41-42

Taiwan, 59
Take-overs, steel, 16ff.
Tariff Schedules of the United States Annotated, 66, 68
Taylor, A. Thomas, 54
Technology, steel:
capital spending, effect on, 11
cost reduction, and, 36ff.
decision to install, 116ff.
Teledyne, Incorporated:
Columbia Tool Steel Company, take-over of, 18
Firth Sterling, Inc., take-over of, 18
Texas Instruments, Inc., 54
"Thin" tinplate, 89, 109
Thornton, Charles B., 54
Time Incorporated, 54
Tin-free steel, 3, 109
Todd Shipbuilding, 106
Tool steel imports:
tonnage of, 60
voluntary limits on, 64, 65-66, 68
Toyota, 102
Trade Expansion Act, 53
Trans-Alaska pipeline, 50-51
Transportation:
aircraft industry, steel demand, 101
automobile industry, steel demand, 3, 100, 101, 103
railroads, steel demand, 101, 103, 104, 105
shipbuilding, steel demand, 101, 105, 107
steel consumption category, as, 100
Truman administration, 35
Truman, Harry S., President, 35
Two-piece can, 3, 110

Union Miniere, 115
United Kingdom, 46, 55, 58, 59, 61, 64, 65, 67, 81
United States Department of Commerce, 64
United States Department of Labor, Bureau of Labor Statistics, 76, 78, 79, 81, 85, 86
United States Department of State, 62
United States Department of the Interior, 106
United States Steel Corporation:
  Acme Steel Company, semi-finished supplied to, 22
  Altos Hornos, interest in, 35
  Baytown, Texas plant of, 33
  Consumers Union suit, named in, 71
  crude steel output, 1970, 114
  direct reduction plant of, 38
  diversification by, 12-13
  formation of, 1901, 114
  iron ore development by, 34
  labor bargaining team, 1971, 75
  largest steel company, position as, 48
  new product introduction by, 84
  plastics business of, 13
  Q-BOP process, installation of, 41
  shipments and employment, 1919 and 1969, 83
  shipments vs. import tonnage, 7, 51
  slab casters installed by, 42
  USS Chemicals Division of, 12-13
United Steelworkers International, 92
United Steelworkers of America:
  aluminum and copper settlements, 1971, 74
  annual convention, 1970, 7
  bargaining team, 1971, 75
  contract negotiations, 1971, 73
  import quotas, steel, support for, 52
Usinor, 114

van der Rest, Pierre, 57, 67
van Veelen, Evert, 57, 68
Venezuela, 38
Vereniging de Nederlandse Ijzer-en Staalproducerende Industria, 57, 68
Veterans' Administration, 98
Volkswagen, 102
Voluntary restraint agreements, steel imports:
  details of, 1968, 54-58
  details of, 1972, 62-70
  opposition to, 1972, 71

Wabush Iron, 34
Wall, John R., 75
Wall Street, steel stocks, and, 16-17
Wall Street Journal, 52
Walter Kidde and Company, Inc., 17
Ward Steel Company, 14
Ward Supply Company, 14
Wankel engine, 95
Watson, Arthur K., 54
Wendel Group, 114
West German steel industry:
  Hamburg, Midland-Ross plant at, 38
  imports from:
    1968-1971, 62
    decline in, 1969, 59
  mergers in, 48
  output of:
    1938, 1947, 1955, 1960, 46
    post-war limits on, 46
  output per man-hour in, 80-81
  productivity comparisons, 87
  voluntary quota proposed by, 54
Wheeling-Pittsburgh Steel Corporation:
  formation of by merger, 23-25
  labor negotiations, 1971, withdrawal from, 74
Wheeling Steel Corporation:
  facility modernization, 1963-1965, 24
  financial difficulties, 1966, 24
  Norton Simon, and:
    interest acquired by, 23
    resignation of, 25
  Operation bootstrap of, 24
  Pittsburgh Steel Company, merger with, 23-25
  PTO-pricing of, 24
Whittaker Corporation, 17
Wilson, Joseph C., 54
Wirtschaftsvereinigung Eisen-und Stahlindustrie, 57, 67
Wisconsin Steel Company, 101
Worthington, L.B., 52
W.R. Grace & Company, 53

Xerox Corporation, 54
Yawata Iron and Steel Company, Ltd., 48, 55, 114, 115
Youngstown Sheet & Tube Company:
  Avnet, Inc., merger talks with, 17
  Erie Mining Company, participation in, 34
  Ling-Temco-Vought, approach by, 17
  Lykes Corporation, merger with, 17, 18
  Signal Corporation, approach by, 17
  take-over bid for, 17, 18, 28